9th International Conference on Mechatronics and Control Engineering (ICMCE 2020)

Held online due to COVID-19

IOP Conference Series: Materials Science and Engineering
Volume 922

Rome, Italy
22 - 24 July 2020

ISBN: 978-1-7138-3167-9
ISSN: 1757-8981

Printed from e-media with permission by:

Curran Associates, Inc.
57 Morehouse Lane
Red Hook, NY 12571

Some format issues inherent in the e-media version may also appear in this print version.

Institute of Physics
Dirac House, Temple Back
Bristol BS1 6BE UK

Phone: 44 1 17 929 7481
Fax: 44 1 17 920 0979

techtracking@iop.org

Additional copies of this publication are available from:

Curran Associates, Inc.
57 Morehouse Lane
Red Hook, NY 12571 USA
Phone: 845-758-0400
Fax: 845-758-2633
Email: curran@proceedings.com
Web: www.proceedings.com

TABLE OF CONTENTS

IOP Conf. Series: Materials Science and Engineering **922** (2020) 011001 doi:10.1088/1757-899X/922/1/011001

PREFACE

It gives me a great pleasure to invite you to the 9th International Conference on Mechatronics and Control Engineering (ICMCE 2020) during July 22-24, 2020 in full virtual mode on ZOOM.

The safety and well-being of participants is always of paramount importance. This year, considering the COVID-19 pandemic and the severe travel restrictions worldwide, after careful consideration and following some participants' advices, the organizing committees decided to convert ICMCE 2020 into a fully virtual conference. It has been a very difficult decision, but it has also given to all of us an opportunity to innovate and to test a new style of conference in a safer and convenient way in this special period. I hope ICMCE 2020 can provide an online platform for every participant to communicate and to share ideas.

Special thanks are extended to our colleagues in the program committee for their thorough review of all the submissions, which is vital to the success of the conference, and also to the members of the organizing committee and to the volunteers who had dedicated their time and efforts in planning, promoting, organizing and helping the conference.

I hope this success can develop into persistent success in future years, involving attendees and speakers from all corners of the globe and all countries. I sincerely hope you have an excellent time during ICMCE 2020.

General Chair of ICMCE 2020

Prof. Giuseppe Conte,

Università Politecnica delle Marche, Italy

Conference Committee

International Advisory Committee

Pierre Borne, IEEE Fellow, Ecole Centrale de Lille, France

Conference General Chair

Giuseppe Conte, Università Politecnica delle Marche, Italy

Conference Co-Chair

Jan Awrejcewicz, The Lodz University of Technology, Poland

Program Chairs

Olivier Sename, Grenoble Institute of Technology, France

Roberto Setola, Campus Biomedico Roma, Italy

Program Co-Chair

Aurelio Piazzi, Università di Parma, Italy

Publicity Chair

Ke-Lin Du, Concordia University, Canada

Technical Committees

Enrico Simetti, Università di Genova, Italy

Antonio Visioli, Universitrà di Brescia, Italy

Elena Zattoni, Universitrà di Bologna, Italy

David Scaradozzi, Universitrà Politecnica delle Marche, Italy

Silvia Maria Zanoli, Universitrà Politecnica delle Marche, Italy

Haijun Gong, Georgia Southern University, USA

Xiaobo Liu-Henke, Ostfalia University of Applied Sciences, Germany

Recai Kuş, Selcuk University, Turkey

Mark Jackson, Kansas State University, USA

Xiaobing Mao, Wuhan University of Technology, China

Sarunya Promkotra, Khon Kaen University, Thailand

Fatih Karpat, Bursa Uludag University, Turkey

Waail Mahmod Lafta, Griffith University, Australia

Zeki Yagiz Bayraktaroglu, Istanbul Technical University, Turkey

Parvaneh Esmaili, Near East University, Cyprus

Emin Taner ELMAS, Igdir University, Turkey

Gokhan Erdemir, Istanbul Sabahattin Zaim University, Turkey

Xiaolin Wang, University of Tasmania, Australia

K. Kadirgama, Universiti Malaysia Pahang, Malaysia

Dario Mendoza Chipantasi, Universidad de las Fuerzas Armadas ESPE, Ecuador

Yimin Zhao, Hofstra University, USA

IOP Conf. Series: Materials Science and Engineering **922** (2020) 011003 doi:10.1088/1757-899X/922/1/011003

Peer review declaration

All papers published in this volume of *IOP Conference Series: Materials Science and Engineering* have been peer reviewed through processes administered by the Editors. Reviews were conducted by expert referees to the professional and scientific standards expected of a proceedings journal published by IOP Publishing.

- **Type of peer review:** Double-blind
- **Conference submission management system:** EasyChair submission system
- **Number of submissions received:** 25 papers
- **Number of submissions sent for review:** 25 papers
- **Number of submissions accepted:** 12 papers
- **Acceptance Rate (Number of Submissions Accepted / Number of Submissions Received X 100):** 48.00%
- **Average number of reviews per paper:** 2
- **Total number of reviewers involved:** 8 reviewers
- **Any additional info on review process:**

  1. Preliminary Review
  2. The papers passed the first review will be reviewed again from the following aspects: Originality, Innovation, Technical Soundness, Applicability, Clarity of presentation and Relevance

- **Contact person for queries:** Jevy Tsang (jevytsang@asr.org)

IOP Conf. Series: Materials Science and Engineering **922** (2020) 012001 doi:10.1088/1757-899X/922/1/012001

Reference tracking of dynamic system using prefilter and integral extension

Jean Oscard Domguia Teto, Jörg Seewig
MTS Institute of Measurement and Sensor-Technology, University of Kaiserslautern, Kaiserslautern, Germany
E-mails: oscard.teto@mv.uni-kl.de, seewig@mv-uni-kl.de

Abstract. In this paper, a control approach for reference tracking in state-space that combines two existing methods for designing two degrees-of-freedom controllers for reference tracking is presented. One of these methods is the use of a prefilter V to scale the reference input in order to compensate the existing DC-Gain of the closed-loop behavior. The second approach incorporates an integral extension in the state-space model by integrating the measured output and embedding it in the dynamic of the system to be controlled. Each of these approaches has advantages as well as disadvantages. By merging these two methods, the goal is to use the advantages of each method to overcome their drawbacks simultaneously.

1. Introduction

Linear control methods belong to the most commonly used control techniques for dynamic system. The biggest advantage of these methods is the simplicity of the design procedures. The first goal the controller is to feed the given dynamic system with an input, that stabilizes the system and make it reach the desired behaviour. However, the mathematical model does not exactly match the dynamic behaviour of the system. To ensure the stability of the system even by variation of some system characteristics, the controller should have a high performance.

Besides the stability of the dynamic system, another relevant aims of a controller is the reference tracking. There exist two well-known approaches for the reference tracking in state-space domain. The first approach divides the controller in two parts. The first part is responsible for stability and the second one ensures a DC-Gain of $k_\infty \neq 1$ between the reference and the output. This structure is not robust again process noises. The second approach used an integral extension of the state-space model. The stability and reference tracking are achieved with a simple state-feedback gain of the new system. This approach is robust against process-noises, but the reaction is slower due to the integrator incorporated in the system.

In this paper, we will first start with some mathematical preliminaries then we will present different approaches for reference tracking. lastly, the effect of each method will be tested on different dynamical systems.

2. Preliminaries

We consider the system dynamic (1) to be controlled. Many methods such as pole-placement and optimal control have been developed to design controllers that fulfil the main task of stability using the state-space representation. If we consider the autonomous part of the system (1) without noise, then the dynamic is given by the eigenvalues of the matrix $\boldsymbol{A}$.

$$\begin{cases} \dot{\boldsymbol{x}} = \boldsymbol{A}\boldsymbol{x} + \boldsymbol{B}u + \boldsymbol{E}d \\ \qquad y = \boldsymbol{C}\boldsymbol{x} \end{cases} . \quad (1)$$

Theorem 2: Consider the autonomous dynamic system $\dot{\boldsymbol{x}} = \boldsymbol{Ax}$. If all the eigenvalues of $\boldsymbol{A}$ are strictly negative, then this system is asymptotically stable.

The aim of the pole-placement procedure is to design a state feedback

$$u = -\boldsymbol{Kx} \;\Rightarrow\; \dot{\boldsymbol{x}} = (\boldsymbol{A} - \boldsymbol{BK})\boldsymbol{x} = \bar{\boldsymbol{A}}\boldsymbol{x} \tag{2}$$

such that, the controlled system has the desired behaviour. [1]

However, all these techniques are based on linear approximations of the real system. To ensure the stability of the real system, the controller should have a high performance. Performance check is easily made in frequency-domain.

Using Laplace-transform, the dynamic of the closed-loop system (Figure 1) can be resumed in

$$Y(s) = G_w(s)W(s) + G_d(s)D(s). \tag{3}$$

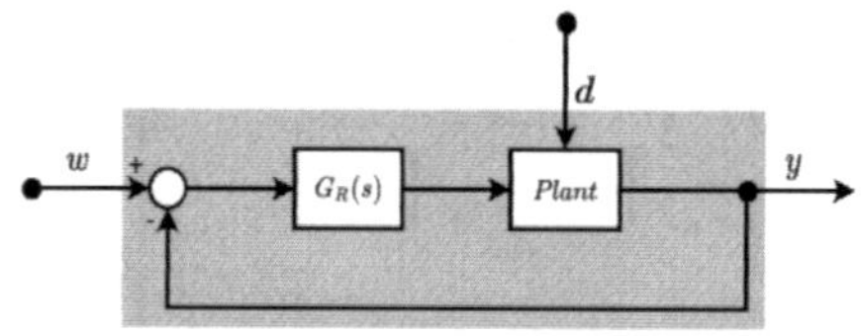

Figure 1. Classical control-loop

To analyse the performance of a controlled system, the closed-loop transfer function for the reference $G_w(s)$ is used. The available tools make used of the open-loop transfer-function $L(s)$, which is used in the well know classical control-loop with negative output feedback (Figure 1).

$$G_w(s) = \frac{Y(s)}{W(s)} = \frac{L}{1+L}. \tag{4}$$

Theorem 3: Nyquist-criteria

An open-loop with the transfer-function $L(s)$ leads to an I/O-stable (input/output) loop if the Nyquist plot $L(\mathrm{j}\omega)$ for $\omega = -\infty \cdots \infty$ encloses the point $-1 + \mathrm{j}0$ of the complex plane $-n^{+}$-time clockwise. Where n^{+} denotes the number of poles of $L(s)$ with positive real part. [2]

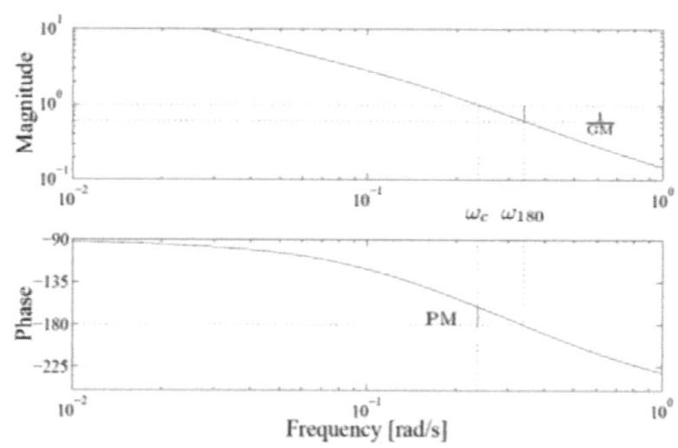

Figure 2. Gain and phase margin in Bode-diagram [3]

Some important frequency-domain measures are used to assess performance e.g. gain and phase margins [3]. Figure 2 shows gain and phase margins in Bode-diagram. The gain margin GM defined as

$$GM = \frac{1}{|L(\mathrm{j}\omega_{180})|} \quad \text{with} \quad \angle L(\mathrm{j}\omega_{180}) = -180°. \text{ [3]} \tag{5}$$

The gain margin is the factor by which the loop gain $|L(\mathrm{j}\omega)|$ can be increased before the closed-loop system becomes unstable. The gain margin is thus a direct safeguard against steady-state gain uncertainty [3]. The phase margin is defined as

$$PM = \angle L(\mathrm{j}\omega_c) + 180°, \quad \text{with} \quad |L(\mathrm{j}\omega_c)| = 1. \text{ [3]} \tag{6}$$

The phase margin tells how much negative phase (phase lag) we can add to $L(s)$ at frequency ω_c before the phase at this frequency becomes $-180°$ which corresponds to closed-loop instability. The phase margin is a direct safeguard against time delay uncertainty [3].

3. Reference Tracking

Reference tracking is a very important task of a controller. In this part, we will present two different approaches for reference tracking in state-space domain. Then we will merge the two approaches.

3.1. Reference Tracking with prefilter

Consider the dynamic System (1) to be controlled with a simple state-feedback matrix and the reference such that

$$u = -\boldsymbol{Kx} + Vw \tag{7}$$

(Figure 3). With the Laplace-transform following relations is derived for reference and noise transfer-functions:

$$G_w(s) = \frac{Y(s)}{W(s)} = \boldsymbol{C}(s\boldsymbol{I} - \boldsymbol{A} + \boldsymbol{BK})^{-1}\boldsymbol{B}V, \quad G_d(s) = \frac{Y(s)}{D(s)} = \boldsymbol{C}(s\boldsymbol{I} - \boldsymbol{A} + \boldsymbol{BK})^{-1}\boldsymbol{E}. \tag{8}$$

For reference tracking, we consider the behaviour of the transfer function for low frequencies ($\omega \to 0$). It holds

$$\lim_{\omega\to\infty} G_w(j\omega) = 0, \qquad \lim_{\omega\to 0} G_w(j\omega) = -\boldsymbol{C}(\boldsymbol{A} - \boldsymbol{BK})^{-1}\boldsymbol{B}V \stackrel{\text{def}}{=} 1 \Rightarrow V = -(\boldsymbol{C}(\boldsymbol{A} - \boldsymbol{BK})^{-1}\boldsymbol{B})^{-1}. \text{ [1]} \quad (9)$$

The behaviour of the noise transfer function $G_d(s)$ should also be analysed to see the effect of process noise on the system.

$$\lim_{\omega\to 0} G_d(j\omega) = -\boldsymbol{C}(\boldsymbol{A} - \boldsymbol{BK})^{-1}\boldsymbol{E}, \ \lim_{\omega\to\infty} G_w(j\omega) = 0 \quad (10)$$

The relation (10) shows that high-frequency noises will be damped, but it will not be the case for low-frequency noises. For instance, step noise will be transferred to the system with the DC-Gain

$$k_{d_\infty} = -\boldsymbol{C}(\boldsymbol{A} - \boldsymbol{BK})^{-1}\boldsymbol{E}.$$

To overcome this problem, another approach is used.

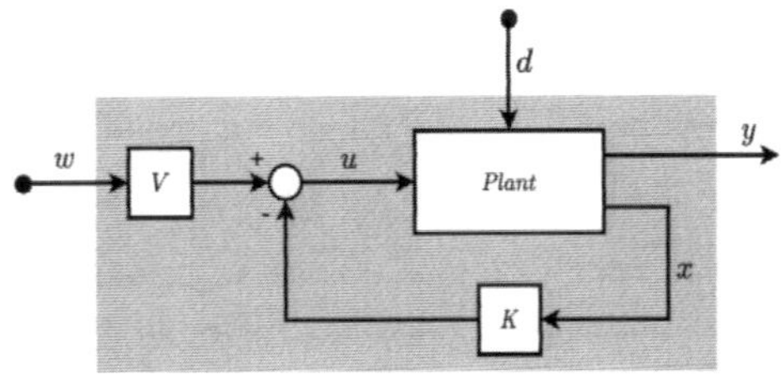

Figure 3. State-feedback with prefilter

3.2. Integral extension for reference tracking

Figure 4 show the idea of the integral extension. In this procedure, the integral of the error $e = y - w$ is included in the state-space model and the new representation is used to design a feedback gain. Consider z as new state-space representation. An equivalent dynamic system can be derived

$$z = \int (y - w)\,dt \;\Rightarrow\; \underbrace{\begin{bmatrix}\dot z\\ \dot{\boldsymbol{x}}\end{bmatrix}}_{\dot{\tilde{x}}} = \underbrace{\begin{bmatrix}0 & \boldsymbol{C}\\ 0 & \boldsymbol{A}\end{bmatrix}}_{\tilde{A}}\underbrace{\begin{bmatrix}z\\ \boldsymbol{x}\end{bmatrix}}_{\tilde{x}} + \underbrace{\begin{bmatrix}0\\ \boldsymbol{B}\end{bmatrix}}_{\tilde{B}}u + \underbrace{\begin{bmatrix}-1\\ \boldsymbol{0}\end{bmatrix}}_{\tilde{E}}w + \underbrace{\begin{bmatrix}0\\ \boldsymbol{E}\end{bmatrix}}_{E_0}d \tag{11}$$

$$y = \underbrace{[0 \quad \boldsymbol{C}]}_{\tilde{C}}\underbrace{\begin{bmatrix}z\\ \boldsymbol{x}\end{bmatrix}}_{\tilde{x}}. \tag{12}$$

Using a state feedback

$$u = -\tilde{\boldsymbol{K}}\tilde{\boldsymbol{x}} = -[K_I \quad \boldsymbol{K_P}]\begin{bmatrix}z\\ \boldsymbol{x}\end{bmatrix} \tag{13}$$

to control the system and applying the Laplace-transform lead to the transfer functions

$$G_w(s) = \frac{Y(s)}{W(s)} = \tilde{\boldsymbol{C}}\left(s\boldsymbol{I} - \tilde{\boldsymbol{A}} + \tilde{\boldsymbol{B}}\tilde{\boldsymbol{K}}\right)^{-1}\tilde{\boldsymbol{E}} = \frac{1}{s}\boldsymbol{C}\left(s\boldsymbol{I} - \boldsymbol{A} + \boldsymbol{BK_P} + \frac{1}{s}\boldsymbol{B}K_I\boldsymbol{C}\right)^{-1}\boldsymbol{B}K_I, \tag{14}$$

$$G_d(s) = \frac{Y(s)}{D(s)} = \tilde{\boldsymbol{C}}\left(s\boldsymbol{I} - \tilde{\boldsymbol{A}} + \tilde{\boldsymbol{B}}\tilde{\boldsymbol{K}}\right)^{-1}\boldsymbol{E}_0 = \boldsymbol{C}\left(s\boldsymbol{I} - \boldsymbol{A} + \boldsymbol{BK_P} + \frac{1}{s}\boldsymbol{B}K_I\boldsymbol{C}\right)^{-1}E. \tag{15}$$

For $\omega \to 0$ and $\omega \to \infty$, it holds

$$\lim_{\omega\to 0} G_w(j\omega) = \boldsymbol{C}(\boldsymbol{B}K_I\boldsymbol{C})^{-1}\boldsymbol{B}K_I \stackrel{(\star)}{\to} 1, \qquad \lim_{\omega\to\infty} G_w(j\omega) = 0, \quad (16)$$

$$\lim_{\omega\to 0} G_d(j\omega) = 0, \qquad \lim_{\omega\to\infty} G_d(j\omega) = 0. \quad (17)$$

($\star$): if $\mathbf{BK_I C}$ non-singular.

For slow varying reference, the tracking occurs successfully. High-frequency noises are damped as well as low-frequency noise. This was not the case with the prefilter. However, due to the integrator embedded in the new dynamic for controller design, the response of the system may be slower.

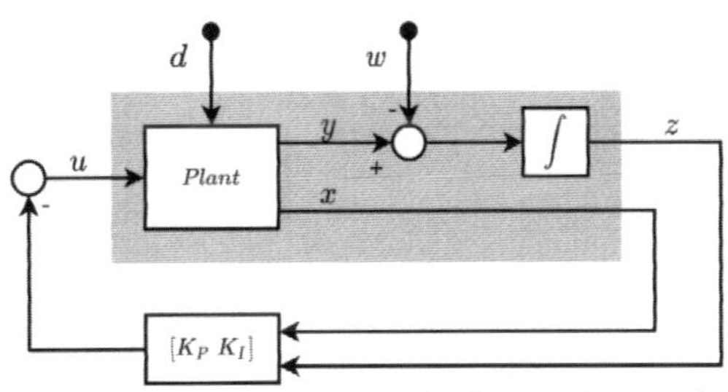

Figure 4. State feedback with integral extension

Now, we will try to combines the two techniques in order to have the fast reaction of the prefilter and the noise rejection of integral extension.

IOP Conf. Series: Materials Science and Engineering **922** (2020) 012001 doi:10.1088/1757-899X/922/1/012001

3.3. Combination of prefilter and integral extension for reference tracking

Now, we consider the extended plant (21) with the control input

$$u = -\widetilde{\boldsymbol{K}}\widetilde{\boldsymbol{x}} + Vw = -[K_I \quad \boldsymbol{K_P}]\begin{bmatrix} z \\ \boldsymbol{x} \end{bmatrix} + Vw, \qquad \text{with} \qquad V = -(\boldsymbol{C}(\boldsymbol{A} - \boldsymbol{BK})^{-1}\boldsymbol{B})^{-1} \quad (18)$$

(Figure 5). The resulting transfer functions are

$$G_w(s) = \boldsymbol{C}\left(s\boldsymbol{I} - \boldsymbol{A} + \boldsymbol{BK_P} + \frac{1}{s}\boldsymbol{B}K_I\boldsymbol{C}\right)^{-1}\boldsymbol{B}\left(\frac{1}{s}\boldsymbol{K_I} + \boldsymbol{V}\right), \quad (19)$$

$$G_d(s) = \boldsymbol{C}\left(s\boldsymbol{I} - \boldsymbol{A} + \boldsymbol{BK_P} + \frac{1}{s}\boldsymbol{B}K_I\boldsymbol{C}\right)^{-1}\boldsymbol{E}. \quad (20)$$

The noise transfer function is the same as from the extended plant. Thus, it is also robust against low and high-frequency noises. For the transfer function $G_w(s)$ we have

$$\lim_{\omega\to 0} G_w(\mathrm{j}\omega) = \boldsymbol{C}(\boldsymbol{B}K_I\boldsymbol{C})^{-1}\boldsymbol{B}K_I \overset{(\star)}{\to} 1 \quad (21)$$

$$\lim_{\omega\to \infty} G_w(\mathrm{j}\omega) = 0. \quad (22)$$

We could not derive some observations about the reference transfer function, but using some examples we want to illustrate it in the next section.

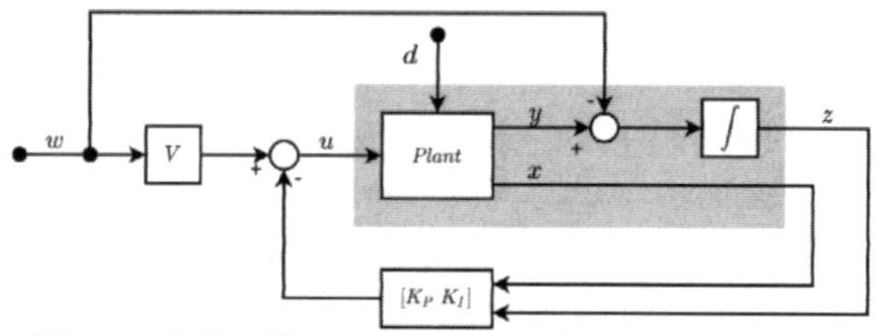

Figure 5. Prefilter combined with state-feedback with integral extension

4. Example

In this section, we want to present the result of each methods of reference tracking control on closed-loop behaviour of a dynamic system. We have tested the three approaches on five different dynamic systems. In this section, one of them will be presented.

The chosen system has following dynamic

$$\dot{x} = \begin{bmatrix} 0 & 1 & 2 \\ 1 & 0 & 1 \\ -2 & -3 & 4 \end{bmatrix} x + \begin{bmatrix} 1 \\ 0 \\ -1 \end{bmatrix} u + \begin{bmatrix} 0 \\ 1 \\ 1 \end{bmatrix} d,$$

$$y = [1 \quad 1 \quad 0]x.$$

For the design of $\boldsymbol{K}$ and V we used

$$p_1 = -2, p_2 = -2 \text{ and } p_3 = -2.5$$

as poles, and for $\widetilde{\boldsymbol{K}} = [K_I \quad \boldsymbol{K_P}]$ we used

$$p_1 = -2, p_2 = -2, p_3 = 2.5 \text{ and } p_4 = -1.5.$$

For each method, we got following transfer functions

$$G_w^{pr}(s) = \frac{-5\,s^2 + 30\,s + 30}{6\,s^3 + 33\,s^2 + 57\,s + 30}$$

$$G_w^{in}(s) = \frac{5.62\,s^2 + 33.78\,s + 33.78}{4.5\,s^4 + 31.53\,s^3 + 79.94\,s^2 + 150}$$

$$G_w^{co}(s) = \frac{-1.13\,s^3 + 5.07\,s^2 + 16.89\,s + 10.13}{1.35\,s^4 + 9.46\,s^3 + 23.98\,s^2 + 26.01\,s + 10.13}$$

^{pr} for prefilter approach, ^{in} for integral extension, ^{co} for combination for the reference and

$$G_d^{pr}(s) = \frac{16\,s^2 + 221\,s + 178}{16\,s^3 + 88\,s^2 + 152\,s + 80}$$

$$G_d^{in}(s) = \frac{4.5\,s^3 + 81.49\,s^2 + 68.26\,s}{4.5\,s^4 + 31.53\,s^3 + 79.94\,s^2 + 86.69\,s + 33.78}$$

$$G_d^{co}(s) = \frac{4.5\,s^3 + 81.49\,s^2 + 68.26\,s}{4.5\,s^4 + 31.53\,s^3 + 79.94\,s^2 + 86.69\,s + 33.78}$$

for the noise. Figure 6 shows the Bode diagram of the reference transfer functions for each approach. For low frequencies, they all have the same magnitude behaviour.

Nevertheless, the prefilter has a higher phase than the integral extension; therefore, it should have a faster reaction and a higher phase margin than the integral extension.
Figure 7 shows the Bode diagram of the noise transfer functions for each method. It shows that the prefilter will transmit low-frequency process noises with an amplification of $6.95\ dB$ while the integral extension will reject them with a higher phase i.e. faster reaction. Both figures show, that from the combination of both approaches applied on the system should result the fast reaction to changes in reference of prefilter and the good noise rejection behaviour of integral extended plant.

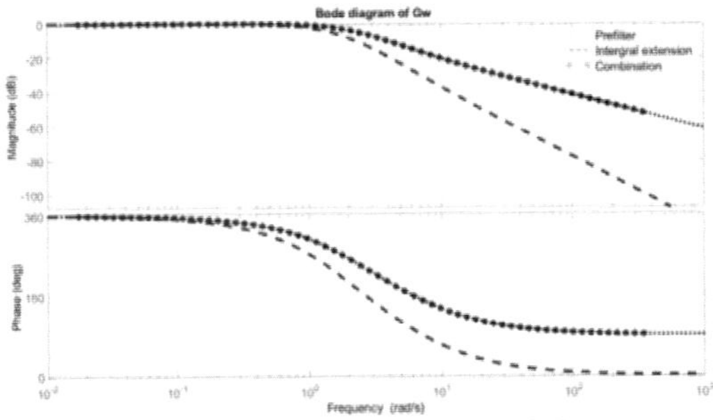

Figure 6. Bode diagram of G_w

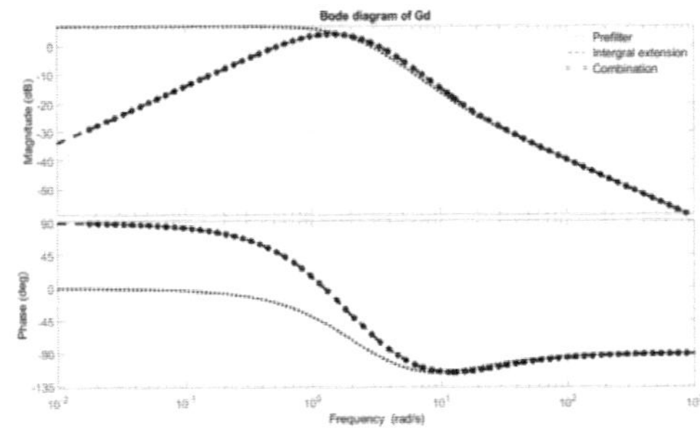

Figure 7. Bode diagram of G_d

The chosen plant was simulated for each controller with the initial state

$$\boldsymbol{x}_0 = [-1 \quad -1 \quad 2]^T$$

and step reference introduced at $t = 5s$ with amplitude 10. A step noise with amplitude 3 was introduced at $t = 10s$. Figure 8 shows on the first graph the output of the system with the prefilter, on the second graph the output of the system with integral extension and on the third graph the output of the system with the combined approach. As expected, the noise was not damped by the prefilter while the integral extension had a slower response. The mixture of prefilter and integral extension has given in this case a faster dynamic than the integral extension and the step noise was successfully rejected.

5. Conclusion

In this paper, we presented two well-known approaches for reference tracking of a control system in time domain using state-space representation and we showed some of their characteristics using the frequency-domain. We first presented the prefilter to compensate the DC-gain of the reference transfer function. The prefilter was not able to reject low-frequency noises, though. Then we presented the integral extension of the state-space model. The integral extension was able to reject the noise, but its reaction for reference tracking is slower. Mixing both approaches eventually yields to a good noise rejection behaviour while increasing the reaction speed of the system. This was tested on five different systems. They all have similar results and one of those examples was presented in Section 4.

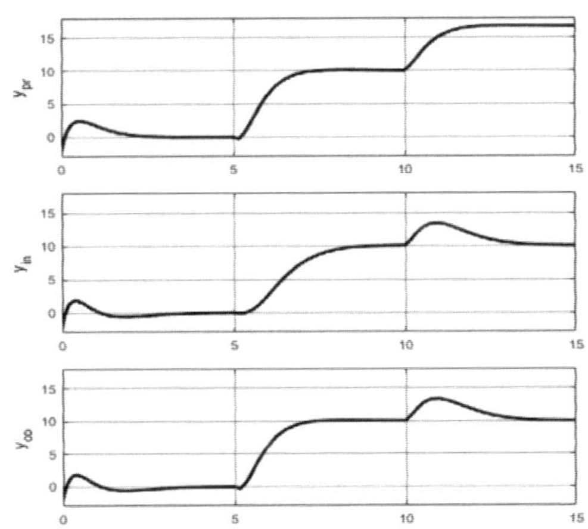

Figure 8. Result of simulation

6. References

[1] Lunze Jan (2016). *Regelungstechnik 2: Mehrgrößensysteme, Digitale Regelung. 9. Auflage*. Springer-Verlag.

[2] Lunze Jan (2013). *Regelungstechnik 1: Systemtheoretische Grundlagen, Analyse und Entwurf einschleifiger Regelungen*. Springer-Verlag.

[3] Sigurd Skogestad, Ian Postlethwaite (2001). *Multivariable Feedback Control Analysis and design.* John Wiley and Sons.

IOP Conf. Series: Materials Science and Engineering **922** (2020) 012002 doi:10.1088/1757-899X/922/1/012002

Invariance and Controlled Invariance in Switching Structured Systems with Application to Disturbance Decoupling

G Conte[1], A M Perdon[1], E Zattoni[2], C H Moog[3] and E Scharbarg[3]

[1] DII, Università Politecnica delle Marche, 60131 Ancona, Italy
[2] DEI "G. Marconi", Alma Mater Studiorum Università di Bologna, 40136 Bologna, Italy
[3] LS2N, UMR CNRS 6004, 44321 Nantes, France

E-mail: {gconte;a.m.perdon}@staff.univpm.it; elena.zattoni@unibo.it; {Claude.Moog; emeric.scharbarg}@ls2n.fr

Abstract. In this paper, we consider dynamical systems where the graph of the relations between state, input and output variables switches between different configurations, according to the action of a switching time signal. Moreover, in each configuration the relations between the variables are known only for being zero or nonzero. Switching structured systems of this kind are described by families of simple, directed graphs. They can be used to model complex networks of systems as well as systems of systems for which the only available information consists in the patterns that the set of the interconnections between the components, or agents, may assume in different situations. Using an approach that is conceptually similar to the geometric approach developed for linear time-invariant systems, suitable notions of invariance and controlled invariance are introduced and related to the action of feedback. These notions are used to provide general solvability conditions for the disturbance decoupling problem expressed in graph-theoretic terms.

1. Introduction

In dealing with dynamical systems, one is often interested in properties that depend on the existence of relations between internal and/or external variables, namely on the structure of the system, rather than in properties that depend on the actual values that specify the relations, namely on the coefficients of the systems. In that case, one speaks of *structured system* and represents them by using simple, directed graphs, whose vertices symbolize the state variables of the system and whose edges symbolize nonzero relations between them. The system properties that are expressed by graph-theoretic characteristics are called *structural* or, since they hold for any value of the nonzero parameters of the system, *generic*. Complex dynamical objects, like networks of systems and systems of systems, are conveniently modeled as structured systems by associating the vertices of the graph to the agents of the network or to the components of the overall system. Thus, features are pointed out that depend on the influence that each agent or component exerts on the others, rather than on the individual dynamics of each of them.
The study of linear structured systems was initiated in [1] and, since then, it has received contributions by many authors. A comprehensive survey of classic results dates back to the early 2000s [2]. How to model and to study complex systems, like networked systems and

systems of systems, by means of structured systems is the object of later papers, like [3], [4], [5], [6], [7], [8], [9], [10], [11] and the references therein.
In a recent paper [12], the authors introduced the notions of controlled invariance and conditioned invariance for a class of structured systems characterized by the fact that each input channel, as well as each output channel, only affects one single vertex of the representing graph. Structured systems of that kind are considered, e.g., in [13], [14], [15], [16], [17] and in some examples of [3]. In the paper mentioned above [12], the novel definition of *essential* state feedback has allowed the notion of controlled invariance for the considered class of structured systems to be characterized in terms of dynamics. Consequently, necessary and sufficient conditions for solvability of the disturbance decoupling problem, either by state or by output feedback, have been established.
Here, we consider structured systems whose structure can vary over the time, thus causing, for instance, a nonzero relation between two variables to vanish or a new relation between previously unrelated variables to appear, as specified by a switching time signal. A system of this kind, that we call *switching structured system*, is modeled by a finite indexed family of structured systems (called *modes*), whose graphs have the same set of vertices, and by a switching time signal, which is assumed to be a piece-wise constant function from $\mathbb{R}^+$ to the set of indices of the family of graphs. Switching structured systems can be used to describe and to study complex dynamical behaviours that are due, for instance, to the action of switches in electrical circuits, to modifications in the topology of communication networks, to variations of the team in multi-agent systems. Examples are given at the end of Section 2.
By introducing a suitable, novel notion of *switching controlled invariance*, we develop a graph-theoretic approach to analysis and synthesis problems for switching structured systems, conceptually akin to the geometric approach for classical linear systems of [19] and [20]. In particular, we show that controlled invariance can be interpreted in terms of feedbacks whose action modifies the modes of the switching structured system at issue. This fact leads to the main theoretic contribution of this paper: the notion of *switching essential state feedback*, defined in graph-theoretic terms.
The developed approach can be used to study a number of control problems that, in particular, include disturbance decoupling problems by means of state feedback. In particular, using controlled invariance, we can state necessary and sufficient solvability conditions that are analogous to those provided in the framework of classical linear systems by the geometric approach. This characterization of the solvability of the disturbance decoupling problem, together with the procedure to construct a solution (if it exists), is the second contribution of this work.
The paper is organized as follows. In Section 2, we introduce the class of switching structured systems considered and we describe their graph representation. In Section 3, we introduce the fundamental notions of switching invariance and switching controlled invariance in graph-theoretic terms. Then, we define the notion of essential switching feedback and we show how it can be used to characterize switching controlled invariance. In Section 4, we study the disturbance decoupling problem by means of state feedback and we characterize its solvability by means of necessary and sufficient conditions. Proofs of our results will appear in complete form elsewhere. Section 5 contains conclusions and description of future work.
Notation Given two sets A and B, we will denote by $A \backslash B$ the set defined by $A \backslash B = \{a \in A, \textit{ such that } a \notin B\}$ and by $A \setminus\!\!\setminus B$ the set defined by $A \setminus\!\!\setminus B = (A \cup B) \backslash (A \cap B)$.

2. Preliminaries
Let $H = \{1, ..., \bar{h}\}$ be a finite set of indices and consider the set $\Gamma = \{(G, \mathcal{E}_h)\}_{h \in H}$, where, for any $h \in H$, $(G, \mathcal{E}_h)$ is a simple directed graph (i.e. a directed graph without multiple edges and without auto-loops) with a finite set of vertices $G = \{v_1, ..., v_n\}$ and set of edges $\mathcal{E}_h \subseteq G \times G$.

Note that all graphs in Γ have the same set of vertices and different sets of edges. If $(v_j, v_i) \in \mathcal{E}_h$, for some $h \in H$, we say that v_j is the tail and v_i is the head of the edge $(v_j, v_i) \in \mathcal{E}_h$. A path P in $(G, \mathcal{E}_h)$ is an ordered finite sequence of edges $(e_1, ..., e_m)$, all belonging to $\mathcal{E}_h$, in which the head of the edge e_k coincides with the tail of the edge e_{k+1}. The tail of the first edge in a path P is called the tail of the path and the head of the last edge is called the head of the path.
To each of the graphs $(G, \mathcal{E}_h) \in \Gamma$, we associate an $n \times n$ matrix $A_h = [a_{ij}]$ whose entries are real, mutually independent parameters that satisfy the following conditions

- $a^h_{ij} \neq 0$ for $i \neq j$ if and only if $(v_j, v_i) \in \mathcal{E}_h$ (i.e. there is an edge from v_j to v_i in $(G, \mathcal{E}_h)$).

Note that no condition is imposed on a^h_{ij} for $i = j$.
Letting $G^{in}_h = \{v_{h1}, ..., v_{hm_h}\} \subseteq G$, for any $h \in H$, be a subset of vertices, we associate to the pair $((G, \mathcal{E}_h), G^{in}_h)$ an $n \times m_h$ matrix $B_h = [b^h_{ij}]$ whose entries are real, mutually independent parameters that satisfy the following conditions

- $b^h_{ij} = 1$ (or more generally $b^h_{ij} \neq 0$) if $v_{hj} = v_i$ (that is: if the j-th element v_{hj} of G^{in}_h is equal to the i-th element v_i of G)
- $b^h_{ij} = 0$ otherwise.

Note that in any column of B_h there is just one entry different from 0, while in any row of B_h there is at most one entry different from 0.
Letting $G^{out}_h = \{v_{h1}, ..., v_{hp_h}\} \subseteq G$, for any $h \in H$, be a subset of vertices, we associate to the pair $((G, \mathcal{E}_h), G^{out}_h)$ a $p_h \times n$ matrix $C_h = [c^h_{ij}]$ whose entries are real, mutually independent parameters that satisfy the following conditions

- $c^h_{ij} = 1$ (or more generally $c^h_{ij} \neq 0$) if $v_{hj} = v_i$ (that is: if the j-th element v_{hj} of G^{out}_h is equal to the i-th element v_i of G)
- $c^h_{ij}j = 0$ otherwise.

Note that in any row of C_h there is just one entry different from 0, while in any column of C_h there is at most one entry different from 0.
In representing graphically the triplet $((G, \mathcal{E}_h), G^{in}_h, G^{out}_h)$ for a given $h \in H$, we use arrows between vertices to indicate edges belonging to $\mathcal{E}_h$ and we mark the elements of G^{in}_h by ingoing arrows and the elements of G^{out}_h by outgoing arrows, as in Figure 1 (where $G = \{v_1, ..., v_5\}$, $\mathcal{E}_h = \{(v_1, v_2), (v_1, v_3), (v_2, v_4), (v_3, v_2), (v_4, v_3), (v_4, v_5)\}$, $G^{in}_h = \{v_1, v_2\}$, $G^{out}_h = \{v_5\}$).
Let $\mathscr{S}$ denote the set of piece-wise constant, left-continuous functions $\sigma : \mathbb{R}^+ \to H$ with a finite number of discontinuities in any interval. The **switching structured system** $\Sigma_\sigma((G, \mathcal{E}_\sigma), G^{in}_\sigma, G^{out}_\sigma))$ associated to the set of triplets $\{(G, \mathcal{E}_h), G^{in}_h, G^{out}_h), h \in H\}$ is the switching linear time-invariant system described in parametric state-space form by the equation

$$\Sigma_\sigma \equiv \left\{ \begin{array}{lcl} \dot{x}(t) & = & A_{\sigma(t)}x(t) + B_{\sigma(t)}u(t) \\ y(t) & = & C_{\sigma(t)}x(t) \end{array} \right. \tag{1}$$

with state $x \in \mathbb{R}^n$, input $u \in \mathbb{R}^m$ and output $y \in \mathbb{R}^p$. The structured linear systems $\Sigma_h \equiv \left\{ \begin{array}{lcl} \dot{x}(t) & = & A_h x(t) + B_h u(t) \\ y(t) & = & C_h x(t) \end{array} \right.$ with $h \in H$ are the modes of the switching structured system Σ_σ and $\sigma(t)$ represents the time signal that governs the switching from one mode to another. Generally, in dealing with structured switching systems, we are interested in features and properties that hold for all $\sigma(t) \in \mathscr{S}$.
Switched structured systems can be used to model complex dynamical structures, like networks of systems or systems of systems, that can switch between different configurations. In that case, the vertices of the graph G represent the systems, or dynamical agents, of the network and the edge $(v_j, v_i) \in \mathcal{E}_h$ represents a link, whose weight is expressed by the parameters a^h_{ij}, that account

IOP Conf. Series: Materials Science and Engineering **922** (2020) 012002 doi:10.1088/1757-899X/922/1/012002

for the way in which the (state of the) agent v_j affects the (dynamics of the) agent v_i when the network is in the configuration corresponding to the index h. The dynamical structure modeled by Σ_σ switches between different configurations according to the time signal $\sigma(t)$ and, as a consequence, some links may disappear and new ones may appear, accounting for a modification of the relationship between agents. The vertices in G_h^{in} and those in G_h^{out} correspond to agents that, respectively, can be influenced by inputs coming from the external environment or that send outputs to the external environment when the system is in the configuration corresponding to the index h.

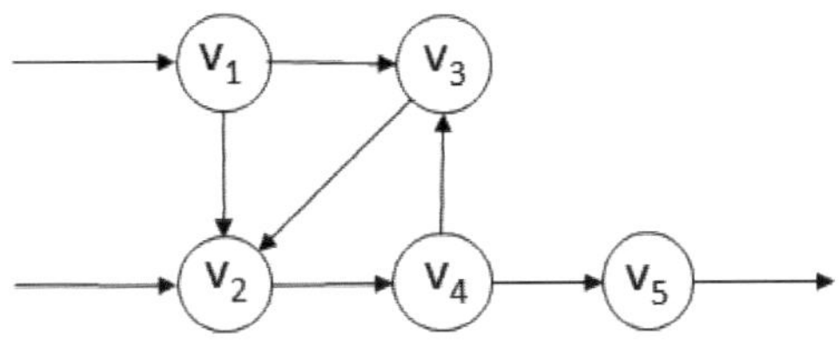

Figure 1: Graphical representation of a triplet $((G, \mathcal{E}_h), G_h^{in}, G_h^{out})$.

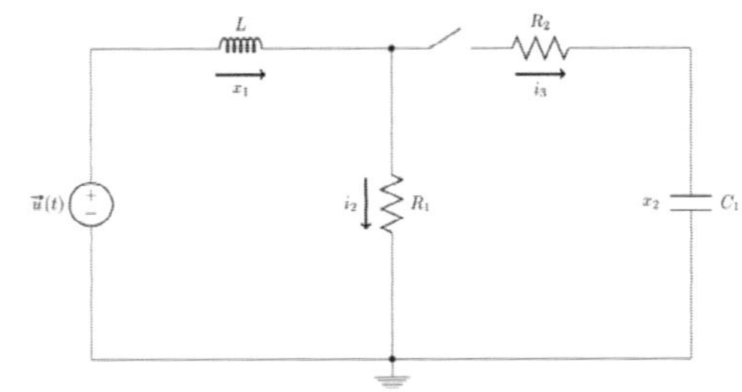

Figure 2: An electrical circuit with one switch.

Example 1 *Consider the electrical circuit of Figure 2, whose configuration varies according to the action of the switch. Choosing the current through the inductor L and the voltage across the capacitor C_1 as the state variables $x_1(t)$ and $x_2(t)$, respectively, and the output $y(t)$ equal to the current through the inductor L if the switch is open and equal to the voltage across the capacitor C_1 if the switch is closed, we get the following equations for the two configurations:*

$$\Sigma_1 \equiv \begin{cases} \dot{x}_1(t) &= \frac{-R_1}{L} x_1(t) + \frac{1}{L} u(t) \\ \dot{x}_2(t) &= 0 \\ y(t) &= x_1(t) \end{cases} \qquad \Sigma_2 \equiv \begin{cases} \dot{x}_1(t) &= \frac{-R_1}{L(R_1+R_2)} (R_2 x_1(t) + x_2(t)) + \frac{1}{L} u(t) \\ \dot{x}_2(t) &= \frac{1}{C_1(R_1+R_2)} (R_1 x_1(t) - x_2(t)) \\ y(t) &= x_2(t) \end{cases}$$

Correspondingly, we have the switching structured system $\Sigma_\sigma((G, \mathcal{E}_\sigma), G_\sigma^{in}, G_\sigma^{in})$, whose two modes have the graphical representation given in Figure 3

Figure 3: Modes of the system $\Sigma_\sigma((G, \mathcal{E}_\sigma), G_\sigma^{in}, G_\sigma^{in})$ that model the circuit of Figure 2.

Example 2 *Consider a cyber-physical system consisting of agents that influence each other by exchanging information through a communication network. If some links of the communication network may be open or closed over a time interval, the system takes different configurations and, without specifying its parameters, it can be conveniently modeled by a switching structured system. Take, for instance, a set of three agents, whose dynamics is one-dimensional, and assume that there are one-directional communication links from the agent v_1 to the agents*

v_2 and v_3 and from the agent v_2 to the agent v_3. Assume that only one of the last two communication links can be disabled at one time, otherwise they are both enabled; that only agent v_1 get inputs from the external environment and only agent v_2 sends outputs to the external environment. Then, the overall system can be described as $\Sigma_\sigma(((G, \mathcal{E}_\sigma), G_h\sigma in, G_\sigma^{out})))$, where $H = \{1,2,3\}$, $G = \{v_1, v_2, v_3\}$, $\mathcal{E}_1 = \{(v_1, v_2), (v_1, v_3), (v_3, v_2)\}$, $\mathcal{E}_2 = \{(v_1, v_2), (v_3, v_2)\}$, $\mathcal{E}_3 = \{(v_1, v_2), (v_1, v_3)\}$, $G_1^{in} = G_2^{in} = G_3^{in} = \{v_1\}$, $G_1^{out} = G_2^{out} = G_3^{out} = \{v_2\}$.The switching signal σ determines if both the communication links between v_1 and v_3 and between v_3 and v_2 are enabled (e.g. $\sigma(t) = 1$), if only the first one is disabled (e.g. $\sigma(t) = 2$) or if only the second one is disabled (e.g. $\sigma(t) = 3$). The triplets $((G, \mathcal{E}_h), G_h^{in}, G_h^{out})$ for $h = 1,2,3$ are represented in Figure 4. The modes Σ_1, Σ_2 and Σ_3 of the systems have, respectively, the following equations:

$$\Sigma_1 \equiv \begin{cases} \dot{x}_1(t) &= a_{11}^1 x_1 + b_{11}^1 u_1(t) \\ \dot{x}_2(t) &= a_{21}^1 x_1 + a_{22}^1 x_2(t) + a_{32}^1 x_3(t) \\ \dot{x}_3(t) &= a_{31}^1 x_1 + a_{33}^1 x_3(t) \\ y(t) &= c_{11}^1 x_3(t) \end{cases}$$

$$\Sigma_2 \equiv \begin{cases} \dot{x}_1(t) &= a_{11}^2 x_1 + b_{11}^2 u_1(t) \\ \dot{x}_2(t) &= a_{21}^2 x_1 + a_{22}^2 x_2(t) + a_{23}^2 x_3(t) \\ \dot{x}_3(t) &= a_{33}^2 x_3(t) \\ y(t) &= c_{11}^2 x_3(t) \end{cases} \qquad \Sigma_3 \equiv \begin{cases} \dot{x}_1(t) &= a_{11}^3 x_1 + b_{11}^3 u_1(t) \\ \dot{x}_2(t) &= a_{21}^3 x_1 + a_{22}^3 x_2(t) \\ \dot{x}_3(t) &= a_{31}^3 x_1 + a_{33}^3 x_3(t) \\ y(t) &= c_{11}^3 x_3(t) \end{cases}$$

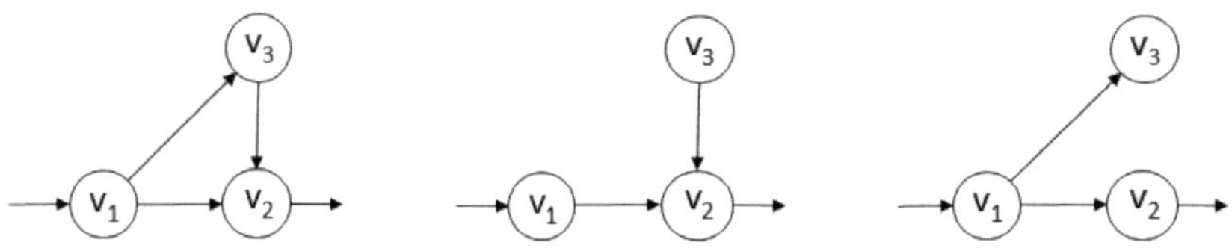

Figure 4: Graphical representation of the triplets $((G, \mathcal{E}_h), G_h^{in}, G_h^{out})$ for $h = 1,2,3$.

Example 3 *The glucose metabolism in the human body can be represented as a switching structured system which is triggered by the external signal consisting of ingestion of carbohydrates as meals and snacks and which aims at reaching a target glycenmic level. As depicted in Figure 5, the core of glucose homeostasis consists in the interactions between the liver and the pancreas. The liver acts a glucose reservoir which releases glucose in blood plasma through some glycogenesis process during fasting periods, like overnight periods. During prandial and postprandial periods, the liver "refills" the glucose reservoir as the full amount of carbohydrates is not consumed in real-time. The pancreas produces essentially either insulin during prandial and postprandial periods, or glucagon during fasting periods to regulate the glucose concentration level in blood plasma. The interaction between the pancreas and the liver can be schematically illustrated as in Figure 6, where the arrows describe how the blood glucose concentration* **G**, *the state of the liver* **L**, *the blood plasma insulinemia* **I**, *the glucagon concentration in the blood plasma* **A**, *the status of the pancreatic alpha cells α and the status of the pancreatic beta cells β influence each other. The input CHO represents the disturbance caused by carbohydrates ingestion. As shown by the basic glucose-insulin-glucagon model of [18], the overall regulation mechanism splits into two actions, which are effective either during the prandial and postprandial periods, when the beta cells produce a higher rate of insulin, or during fasting periods, when alpha cells produce glucagon to stimulate glycogen release by the liver. We have therefore the two modes graphically described in Figure 7: in case of hypoglycemia (G is below the target glycemic level), the glucagon secretion is maximal, with an insulin secretion close to zero (Figure 7 on the left); in case of hyperglycemia (G exceeds the target glycemic level) after ingestion of carbohydrates, the insulin secretion will be maximal, with a glucagon secretion close to zero (Figure 7 on the right).*

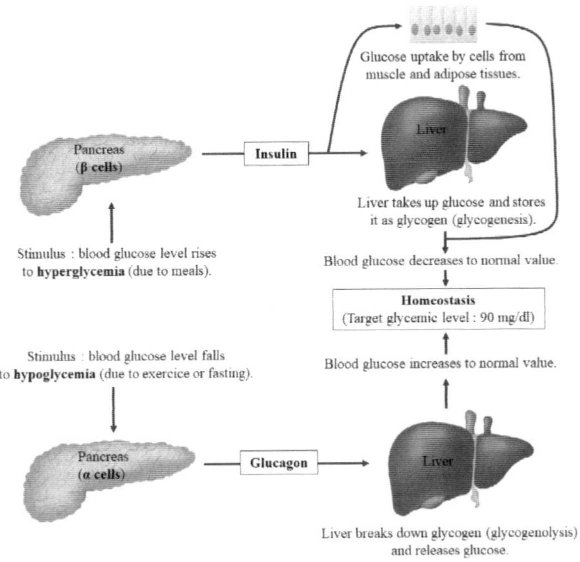

Figure 5: Glucose metabolism.

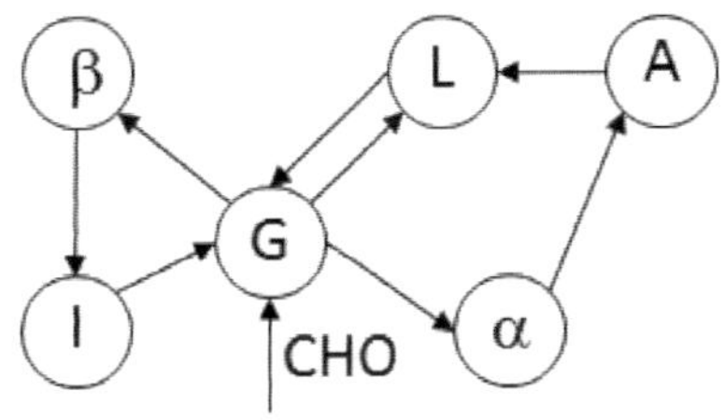

Figure 6: Functional scheme of the glucose metabolism.

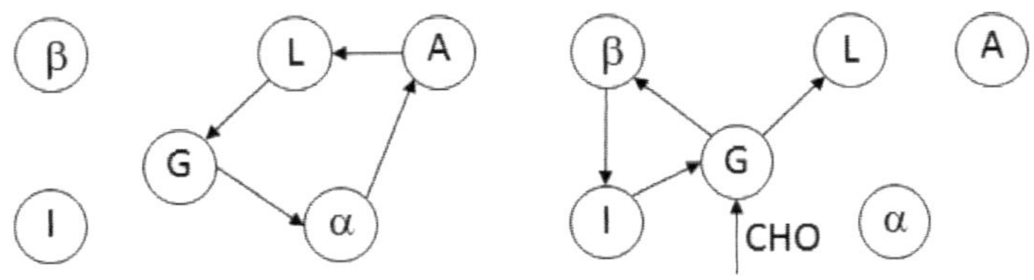

Figure 7: Modes of the system representing the glucose metabolism: hypoglycemia case (left) and hyperglycemia case (right).

3. Invariance and controlled invariance

In this section, we extend to switching structured system the geometric notions that, in accordance with the approach developed for linear systems in [19] and [20], characterize invariance and controlled invariance for structured system (see [2], [12]).

In order to proceed, we introduce the following notions. Given a switching time signal $\sigma(t) \in \mathscr{S}$, we denote by $|\sigma(t)| = (h_1, h_2, ...)$, with $h_k \in H$ and $h_k \neq h_{k+1}$, the ordered sequence of different values that are assumed by $\sigma(t)$ as t goes from 0 to $+\infty$. In practice, $|\sigma(t)|$ describes the sequence of modes that the system $\Sigma_\sigma((G, \mathcal{E}_\sigma), G_\sigma^{in}, G_\sigma^{out})$ goes across when the implemented switching time signal is $\sigma(t)$.

Definition 1 *Given a structured switching system $\Sigma_\sigma((G, \mathcal{E}_\sigma), G_\sigma^{in}, G_\sigma^{out})$, a* **switching path** *P_s in $(G, \mathcal{E}_\sigma)$ is an ordered finite sequence of edges $(e_{h_1}, e_{h_2}, ..., e_{h_m}) \in \mathcal{E}_{h_1} \times ... \times \mathcal{E}_{h_m}$, in which the head of the edge e_{h_k} coincides with the tail of the edge $e_{h_{k+1}}$. The tail of the first edge in a switching path P_s is called the tail of the switching path and the head of the last edge is called the head of the switching path.*

A switching path P_s in $(G, \mathcal{E}_\sigma)$, with $P_s = (e_{h_1}, e_{h_2}, ..., e_{h_m}) \in \mathcal{E}_{h_1} \times ... \times \mathcal{E}_{h_m}$, indicates that the agent represented by the tail of P_s, say v_1, influences the dynamics of the agent represented by the head of P_s, say v_m, if the switching of Σ_σ is governed by a time signal $\bar{\sigma}$ such that $|\bar{\sigma}| = (..., h_1, h_2, ..., h_m, ...)$.

In a way conceptually similar to that followed in [12], we can introduce in our framework a suitable notions of **switching invariance** and **switching controlled invariance** as follows.

Definition 2 *Given a switching structured system* $\Sigma_\sigma((G,\mathcal{E}_\sigma),G_\sigma^{in},G_\sigma^{out})$*, a subset* $V \subseteq G$ *of vertices is said to be* **switching invariant** *for* Σ_σ *if, for any* $h \in H$*,* $(v_j, v_i) \in \mathcal{E}_h$ *with* $v_j \in V$ *implies* $v_i \in V$*.*

Definition 3 *Given a structured system* $\Sigma_\sigma((G,\mathcal{E}_\sigma),G_\sigma^{in},G_\sigma^{out})$*, a subset* $V \subseteq G$ *of vertices is said to be* **switching controlled invariant** *for* Σ_σ *if, for any* $h \in H$*,* $(v_j, v_i) \in \mathcal{E}_h$ *with* $v_j \in V$ *implies* $v_i \in V \cup G_h^{in}$*.*

The structural notion of controlled invariance can be characterized in dynamical terms by using a suitable notion of feedbacks. Following the approach developed in [12], we focus on switching state feedbacks that, acting on each mode of a given switching structured system $\Sigma_\sigma(G,\mathcal{E}_h),G_h^{in},G_h^{out})$, modify the dynamics expressed by its underlying graph. Remarking that the action of a state feedback on a mode $\Sigma_h((G,\mathcal{E}_h),G_h^{in},G_h^{out})$ may only involve the relationship between any component of the state, i.e any element of G, and those components whose dynamics is directly affected by the inputs, i.e any element of G_h^{in}, we recall the definition below from [12].

Definition 4 *Given the structured system* $\Sigma_h((G,\mathcal{E}_h),G_h^{in},G_h^{out})$*, an* **essential state feedback** *consists of a subset* $\mathcal{F}_h \subset G \times G_h^{in}$ *such that* $(v_j, v_i) \in \mathcal{F}_h$ *implies* $i \neq j$*. The action of an essential state feedback* $\mathcal{F}_h$ *on the structured system* $\Sigma_h((G,\mathcal{E}_h),G_h^{in},G_h^{out})$ *gives rise to the compensated switching structured system* $\Sigma_h^{comp}((G,\mathcal{E}_h \setminus\!\setminus \mathcal{F}_h),G_h^{in},G_h^{out})$*, where* $\mathcal{E}_h \setminus\!\setminus \mathcal{F}_h = (\mathcal{E}_h \cup \mathcal{F}_h) \setminus (\mathcal{E}_h \cap \mathcal{F}_h)$*.*

Now, we can give the following novel definition.

Definition 5 *Given the switching structured system* $\Sigma_\sigma(G,\mathcal{E}_\sigma),G_\sigma^{in},G_\sigma^{out})$*, an* **essential switching state feedback** $\mathcal{F}_\sigma$ *consists of a set* $\{\mathcal{F}_h\}_{h \in H}$*, where* $\mathcal{F}_h$ *is an essential state feedback for* $\Sigma_h((G,\mathcal{E}_h),G_h^{in},G_h^{out})$ *for any* $h \in H$*.*

Note that applying an essential switching state feedback $\mathcal{F}_\sigma$ to $\Sigma_\sigma((G,\mathcal{E}_\sigma),G_\sigma^{in},G_\sigma^{out})$ means to modify $\mathcal{E}_h$ (unless $\mathcal{F}_h = \{\emptyset\}$), either by adding new elements of the form (v_j, v_i) with $v_i \in G_h^{in}$ to it or by removing elements of the same form, if present, from it. This modifies the set of switching paths and hence the way in which agents influences each other as the system switches. For any $h \in H$, we associate to the essential switching state feeback $\mathcal{F}_h$ a linear relation of the form $u = F_h x$, where $F_h = [f_{ij}^h]$ is an $m_h \times n$ matrix whose entries are real, mutually independent parameters that satisfy specific conditions as described below. First, let us remark that if (v_j, v_i) belongs to $\mathcal{E}_h \cap \mathcal{F}_h$, then we have that v_i is, for some k, the k-th element of G_h^{in}, i.e. $v_i = v_{hk} \in G_h^{in}$, and, hence, $b_{ik}^h \neq 0$ in the matrix B_h. Now, let us take the parameters f_{ij}^h as follows

- $f_{ij}^h \neq 0$ if and only if $(v_j, v_i) \in \mathcal{F}_h$
- $f_{ij}^h = -a_{ij}^h / b_{ik}^h$ if $(v_j, v_i) \in \mathcal{F}_h$ with $v_i = v_{i_k} \in G_h^{in}$ and a_{ij}^h is different from 0.

Note that no condition is imposed on f_{ij}^h if $i = j$ and if, for $i \neq j$ and $(v_j, v_i) \in \mathcal{F}_h$ with $v_i = v_{hk} \in G_h^{in}$, one has $a_{ik}^h = 0$.
With the above choice, the compensated system $\Sigma_\sigma^{comp}((G,\mathcal{E}_\sigma \setminus\!\setminus \mathcal{F}_\sigma),G_\sigma^{in},G_\sigma^{out})$ turns out to be defined in parametric form by the following set of equations

$$\Sigma_\sigma^{comp} \equiv \begin{cases} \dot{x}(t) &= (A_\sigma + B_\sigma F_\sigma)x(t) + B_\sigma u(t) \\ y(t) &= C_\sigma x(t) \end{cases} \tag{2}$$

Example 4 *In the same vein of Example 2, let us consider the switching structured system* $\Sigma_\sigma(G,\mathcal{E}_\sigma),G_\sigma^{in},G_\sigma^{out})$*, where* $H = \{1, 2\}$*, whose modes are defined by the triplets represented in Figure 8.By applying the switching essential state feedback* $\mathcal{F}_\sigma$ *with* $\mathcal{F}_1 = \{(v_3, v_1)\}$ *and* $\mathcal{F}_2 = \{(v_4, v_1)\}$*, we obtain the compensated system* $\Sigma_\sigma^{comp}(G,\mathcal{E}_\sigma \setminus\!\setminus \mathcal{F}_\sigma),G_\sigma^{in},G_\sigma^{out})$ *whose modes are defined by the triplets represented in Figure 9.*

IOP Conf. Series: Materials Science and Engineering **922** (2020) 012002 doi:10.1088/1757-899X/922/1/012002

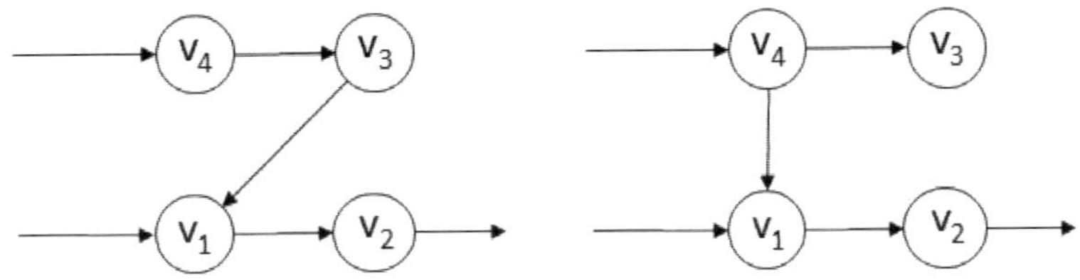

Figure 8: Graphical representation of the triplets $((G, \mathcal{E}_h), G_h^{in}, G_h^{out})$ for $h = 1, 2$.

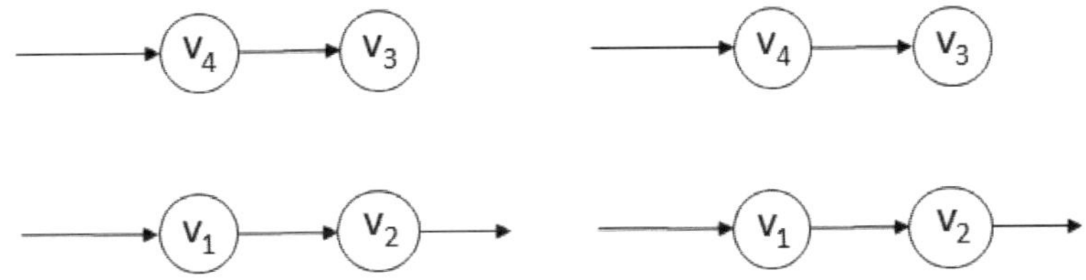

Figure 9: Graphical representation of the triplets $((G, \mathcal{E}_h \setminus\setminus \mathcal{F}_h), G_h^{in}, G_h^{out})$ for $h = 1, 2$.

We can now give a basic dynamical characterization of switching controlled invariance that turns out to be formally analogous to the one given for the corresponding notion in the classical linear framework (compare with [19], [20]).

Proposition 1 *Given a switching structured system $\Sigma_\sigma((G, \mathcal{E}_\sigma), G_\sigma^{in}, G_\sigma^{out})$, a subset of vertices $V \subseteq G$ is switching controlled invariant for Σ_σ if and only if there exists an essential switching state feedback $\mathcal{F}_\sigma$ such that V is switching invariant for the structured switching compensated system $\Sigma_\sigma^{comp}((G, \mathcal{E}_\sigma \setminus\setminus \mathcal{F}_\sigma), G_\sigma^{in}, G_\sigma^{out})$.*

Any essential switching state feedback $\mathcal{F}_\sigma$ that has the property of making V switching invariant in $\Sigma_\sigma^{comp}((G, \mathcal{E}_\sigma \setminus\setminus \mathcal{F}_\sigma), G_\sigma^{in}, G_\sigma^{out})$ is called a **friend** of V.
As switching controlled invariant subspaces in the framework of linear systems, switching controlled invariant subsets of vertices form a semi-lattice with respect to union and inclusion of sets. This property implies the result of the following proposition.

Proposition 2 *Given a switching structured system $\Sigma_\sigma((G, \mathcal{E}_\sigma), G_\sigma^{in}, G_\sigma^{out})$ and a subset $K \subseteq G$, there exists a maximal subset of vertices V such that $V \subseteq K$ and V is switching controlled invariant for Σ_σ. We denote such subset by $V^*(\mathcal{E}_\sigma, G_\sigma^{in}, K)$ or simply by V^* if no confusion arises.*

Given a structured switching system $\Sigma_\sigma((G, \mathcal{E}_\sigma), G_\sigma^{in}, G_\sigma^{out})$ and a subset $K \subseteq G$, it is possible to construct $V^*(\mathcal{E}_\sigma, G_\sigma^{in}, K)$ by considering the sequence of subset $V_k \subseteq G$ defined recursively by

$$\begin{array}{rcl} V_0 & = & K \\ V_{k+1} & = & V_k \setminus \bigcup_{h \in H} \{v_j, \ such\ that\ (v_j, v_i) \in \mathcal{E}_h \ and\ v_i \notin V_k \cup G_h^{in}\}. \end{array} \tag{3}$$

The sequence V_k converges to $V^*(\mathcal{E}_\sigma, G_\sigma^{in}, K)$ in at most $r = card(K)$ steps.

4. Disturbance decoupling by state feedback

A structured switching system subject to disturbance is a system $\Sigma_\sigma((G, \mathcal{E}_\sigma), G_\sigma^{in}, G_\sigma^{out})$ in which G_h^{in} is partitioned as $G_h^{in} = G_h^c \cup G_h^d$ (possibly with $G_c \cap G_d = \emptyset$) for any $h \in H$. Interpreting Σ_σ as a network of agents, G_h^d describes the set of agents that are influenced by a disturbance input when the network is in the configuration corresponding to the index h, while, in the same situation, G^c_h describes the set of agents that are influenced by a control input. For a structured switching system subject to disturbance we can consider the problem of finding a feedback such that, in the compensated system, the disturbance cannot influence the output. In order to analyze this problem, let us remark that the disturbance input influences the output if and only if there is a switching path P_s in $(G, \mathcal{E}_\sigma)$ with tail in G^d_h and head in $G^{out}_{h'}$ for some $h, h' \in H$. Accordingly, we state the disturbance decoupling problem in the following way.

Problem 1 *Given a disturbed structured switching ystem* $\Sigma_\sigma((G, \mathcal{E}_\sigma), (G^c_\sigma \cup G^d_\sigma), G^{out}_\sigma)$, *the* **Disturbance Decoupling Problem by State Feedback** *(DDPSF) consists in finding an essential switching state feedback* $\mathcal{F}_\sigma$, *if any exists, such that in the compensated system* $\Sigma^{comp}_\sigma((G, \mathcal{E}_\sigma \setminus\!\!\setminus \mathcal{F}_\sigma), (G^c_\sigma \cup G^d_\sigma), G^{out}_\sigma)$ *there are no switching paths in* $(G, \mathcal{E}_\sigma \setminus\!\!\setminus \mathcal{F}_\sigma)$ *with tail in* G^d_h *and head in* $G^{out}_{h'}$ *for any* $h, h' \in H$.

Applying the procedure that has been used to derive the system of equations (1), we get the following representation in parametric terms of the disturbed system $\Sigma_\sigma((G, \mathcal{E}_\sigma), (G^c_\sigma \cup G^d_\sigma), G^{out}_\sigma)$

$$\Sigma_\sigma \equiv \begin{cases} \dot{x}(t) &= A_\sigma x(t) + B_\sigma u(t) + D_\sigma d(t) \\ y(t) &= C_\sigma x(t) \end{cases} \tag{4}$$

where, the matrices $B_h = [b^h_{ij}]$ and $D_h = [d^h_{ij}]$ have dimensions, respectively, $n \times c_h$ and $n \times d_h$, with $c_h = card(G^c_h)$ and $d_h = card(G^d_h)$ and, letting $G^c_h = \{v_{h1}, ..., v_{hc_h}\}$ and $G^d_h = \{v_{h1}, ..., v_{hd_h}\}$, their entries satisfies the following conditions

- $b^h_{ij} = 1$ (more generally $b_{ij} \neq 0$) if $v_{hj} = v_i \in$ (i.e. the j-th element of G^c_h is equal to v_i)
- $d_{ij} = 1$ (more generally $b_{ij} \neq 0$) if $v_{hj} = v_i \in$ (i.e. the j-th element of G^d_h is equal to v_i)
- $b_{ij} = 0$ and $d_{ij} = 0$ otherwise

with control input $u \in \mathbb{R}^{c_h}$ and disturbance input $d \in \mathbb{R}^{d_h}$.
Solvability of the DDPSF stated above in the framework of structured switching sytems means solvability of the problem for all values of the parameters which appear in (4). The following theorem gives necessary and sufficient condition for the solvability of the DDPSF and it indicates how to construct a switching feedback $\mathcal{F}_\sigma$ that solves it, if any exists.

Theorem 1 *Given a structured switching disturbed system* $\Sigma_\sigma((G, \mathcal{E}_\sigma), (G^c_\sigma \cup G^d_\sigma), G^{out}_\sigma)$, *the associated DDPSF is solvable if and only if the condition* $G^d_h \subseteq V^*(\mathcal{E}_\sigma, G^c_\sigma, G \setminus \cup_{i \in I} G^{out}_i)$ *is satisfied for all* $h \in H$.

Hint of proof *Sufficiency.* Let $\mathcal{F}_\sigma$ be a friend of $V^*(\mathcal{E}_\sigma, G^c_\sigma, G \setminus \cup_{i \in I} G^{out}_i)$, so that $V^*(\mathcal{E}_\sigma, G^c_\sigma, G \setminus \cup_{i \in I} G^{out}_i)$ is switching invariant in $\Sigma((G, \mathcal{E}_\sigma \setminus\!\!\setminus \mathcal{F}_\sigma), (G^c_\sigma \cup G^d_\sigma), G^{out}_\sigma)$. The above condition implies that, for any $h \in H$, any edge in $\mathcal{E}_\sigma \setminus\!\!\setminus \mathcal{F}_\sigma$ with tail in G^d_h has its head in $V^*(\mathcal{E}_\sigma, G^c_\sigma, G \setminus \cup_{i \in I} G^{out}_i)$ and, therefore, in $G \setminus \cup_{i \in I} G^{out}_i$. This implies that there are no switching paths in $(G, \mathcal{E}_\sigma \setminus\!\!\setminus \mathcal{F}_\sigma)$ with tail in G^d_h and head in G'^{out}_h for any $h, h' \in H$.
Necessity. Let $\mathcal{F}_\sigma$ be a solution of the DDPSF and consider the largest switching invariant $V \subseteq G \setminus \cup_{i \in I} G^{out}_i$ for the compensated system $\Sigma((G, \mathcal{E}_\sigma \setminus\!\!\setminus \mathcal{F}_\sigma), (G^c_\sigma \cup G^d_\sigma), G^{out}_\sigma)$. Clearly, $G^d_h \subseteq V$ for all $h \in H$ and V is switching controlled invariant for $\Sigma_\sigma((G, \mathcal{E}_\sigma), (G^c_\sigma \cup G^d_\sigma), G^{out}_\sigma)$. The conclusion follows by maximality of $V^*(\mathcal{E}_\sigma, G^c_\sigma, G \setminus \cup_{i \in I} G^{out}_i)$. □

Example 5 *Consider the simple example provided by the disturbed structured switching systems* $\Sigma_\sigma((G, \mathcal{E}_\sigma), (G^c_\sigma \cup G^d_\sigma), G^{out}_\sigma)$, *with* $H = \{1, 2\}$, *whose modes are defined by the triplets represented in Figure 10. Computations show that* $G^d_h \subseteq V^*(\mathcal{E}_\sigma, G^c_\sigma, G \setminus \cup_{i \in I} G^{out}_i)$. *Therefore, the condition of Theorem 1 is satisfied. The switching state feedback* $\mathscr{F}\sigma = \{F_1, F_2\}$, *with* $F_1 = \{(v_1, v_c\}$ *and* $F_2 = \{(v_d, v_c\}$, *gives rise to the compensated systems* $\Sigma^{comp}_\sigma((G, \mathcal{E}_\sigma \setminus\!\!\setminus \mathcal{F}_\sigma), (G^c_\sigma \cup G^d_\sigma), G^{out}_\sigma)$, *whose modes are defined by the triplets represented in Figure 11 and it solves the DDPSF.*

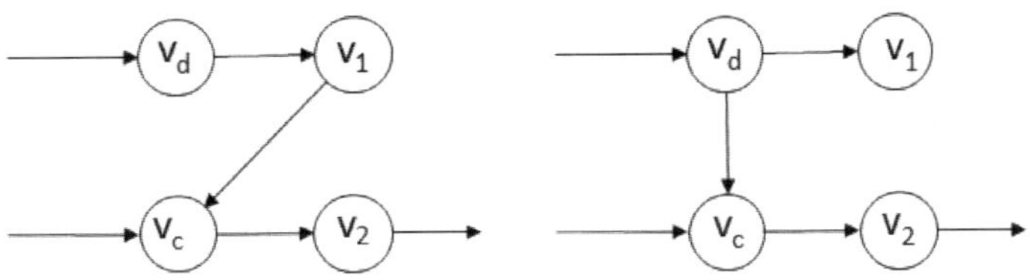

Figure 10: The modes of $\Sigma_\sigma((G,\mathcal{E}_\sigma),(G^c_\sigma \cup G^d_\sigma),G^{out}_\sigma)$ for $h=1,2$.

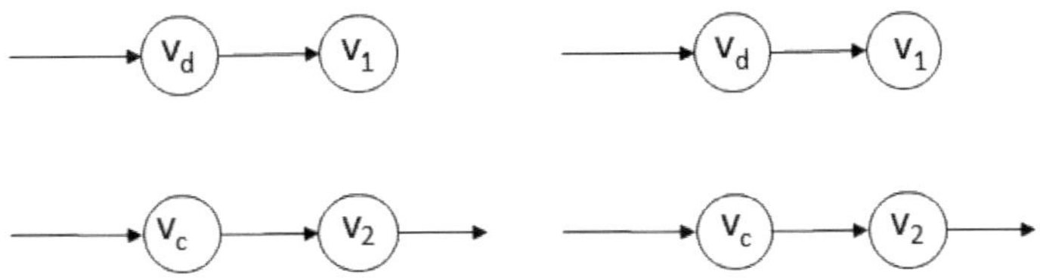

Figure 11: The modes of $\Sigma^{comp}_\sigma((G,\mathcal{E}_\sigma \setminus\setminus \mathcal{F}_\sigma),(G^c_\sigma \cup G^d_\sigma),G^{out}_\sigma)$ for $h=1,2$.

5. Conclusions

A novel approach based on the notion of controlled invariance and of essential feedbacks has been developed for the class of structured systems whose structure varies according to a switching time signal. This approach makes possible to characterize solvability of classical disturbance decoupling problems and to construct solutions. Future work will aim at exploiting this approach in other non-interacting control problems for switching structured systems and in developing specific applications to complex networked systems and systems of systems.

References

[1] Lin C 1974 *IEEE TAC* **19** 35—41
[2] Dion J, Commault C and van der Woude J 2003 *Automatica* **39** 1125–1144
[3] Mesbahi M and Egerstedt M 2010 *Graph Theoretic Methods in Multiagent Networks* (Princeton: Princeton University Press)
[4] Chapman A and Mesbahi M 2013 *Proc. American Control Conference* 6126—6131
[5] Gao J, Liu Y, D'Souza R and Barabasi A 2014 *Nature Communications* **5**
[6] Jia J, van Waarde H, Trentelman H and Camlibel M 2020 *IEEE TAC* early access article
[7] Liu Y and Barabasi A 2016 *Rev. Mod. Phy* **88** 201–208
[8] Monshizadeh N, Zhang S and Camlibel M 2015 *Systems& Control Letters* **76**
[9] Rahmani A, Ji M, Mesbahi M and Egerstedt M. 2009 *SIAM J. Control Optim.* **48** 162–186
[10] Rapisarda P, Everts A and Camlibel M 2015 *Proc. 54th IEEE CDC* 3816–3821
[11] Harrison W 2016 *IEEE Access* **4** 1716–1742
[12] Conte G, Perdon A, Zattoni E and Moog C 2019 *IOP Conf. Series: Materials Science and Engineering* **707**
[13] van Waarde H, Monshizadeh N, Trentelman H and Camlibel M 2019 *Structural Methods in the Study of Complex Systems* (Springer Lectures Notes in Control and Information Science vol 482) ed E Zattoni, A M Perdon et al. 91–112
[14] Trefois M and Delvenne J 2015 *Linear Algebra and its A lications* **484** 199–218.
[15] Monshizadeh N, Zhang S and Camlibel M.K. 2014 *IEEE TAC* **59** 2562–2567
[16] Burgarth D, D'Alessandro D, Hogben L, Severini S and Young M 2013 *IEEE TAC* **58** 2349–2354
[17] Califano C, Scharbarg E, Magdelaine N and Moog C 2019 *Annual Reviews in Control* **48** 233–241
[18] Adams C and Lasseigne D 2018 *Letters in Biomathematics* **5** 70–90
[19] Basile G and Marro G 1992 *Controlled and Conditioned Invariants in Linear System Theory*(Englewood Clffs: Prentice Hall)
[20] Wonham M 1985 *Linear Multivariable Control: A Geometric A roach* (New York: Springer-Verlag)

IOP Conf. Series: Materials Science and Engineering **922** (2020) 012003 doi:10.1088/1757-899X/922/1/012003

Development of high-performance safety controller for excavator

Ka An[1,2]

[1]Jiangsu XCMG Construction Machinery Research Institute LTD
[2]Tuolanshan Road No.26, Economic Development District, Xuzhou, China

Abstract. This article describes the development of one high-performance safety controller for excavators that fulfils SIL2 level complied with IEC 61508 and PL d, Cat.3 level complied with EN ISO 13849. This controller has two microcontrollers, one for standard function and the other for real time diagnostic via the cross communication, it include various functionalities like digital input, digital output, analogue input and PWM output. The embedded software is divided into two parts: safety related part and non-safety related part. These two parts are totally isolated to make sure they will not affect each other. This will protect the operators and vehicles form damage because of lack of practical experience or program fleet.

1. Introduction

Excavator is a kind of earth moving machine, which uses bucket to transfer materials, its main operation objects are soil and rock. From the development of construction machinery in recent years, the excavator market has grown rapidly and has become one of the most important construction markets. Excavator controller is the brain of the whole machine; it receives various sensor signals and drives different types of actuators.

Functional safety is one of most important requirements for machinery in nowadays market. The development trend of digitalization and automation of machinery brings new opportunities and challenges to the functional safety technology. Machinery must be protected against the risks related to function. The safety function of machinery or its control system should ensure the normal realization of the function no matter in normal or fault conditions. Functional safety is also very important for construction machinery to comply with European standards, which is closely related to the high-end market demand of developed countries. IEC 61508[1] and ISO 13849[2] are both related to European market access and are important international standards issued by the international standards committee. If the product fails to meet the requirements of the standards, it will be rejected to enter the high-end market in Europe and America [3].

Therefore, it is of great significance to develop one excavator controller to break through the key technology of functional safety software and hardware design.

2. Hardware design

This high-performance functional safety controller have two microcontrollers, the main chip is TMS570 from Texas Instruments [4]. This device is specifically designed to fulfil safety requirements up to PL e and SIL3. To fulfil the requirements for Category 3 according to ISO 13849-1[5], it is necessary to add an additional microcontroller as companion chip and to switch off the whole system in case of a detected failure, here we choose STM from ST as the companion chip.

IOP Conf. Series: Materials Science and Engineering **922** (2020) 012003 doi:10.1088/1757-899X/922/1/012003

The TMS570 is the only processor that processes the safety function. The STM is doing additional needed diagnostics but not running the safety function itself. According to RFU CNB/M/11.059, the STM system will be developed according to SIL1 requirements.

As Figure 1 shows, power supply is cross monitored, analogue input and digital input signals are detected to both main MCU and companion MCU, then these signals are compared through the communication between two MCUs, about the output, all the output channels have two shut off path.

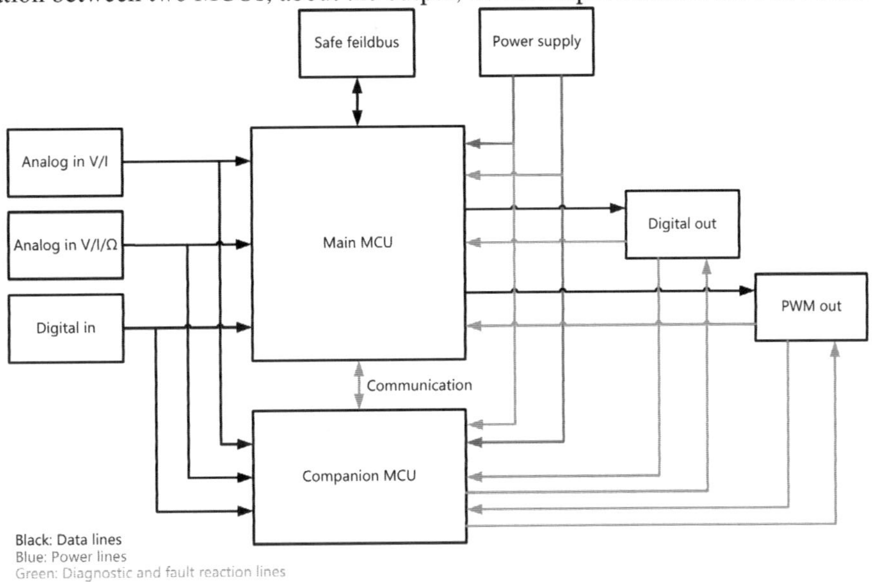

Figure 1. Hardware architecture overview

The external power supply is expected to be +24V, the voltage value will be converted to +12V, +5V, +3.3V, +1.2V on board, different voltages supply power to different parts or components. All of the power supply voltages are monitored by TMS570, STM32 and discrete circuit, when a failure is detected by any of the three paths, the power supply will be switched off. This solution does not feature any separation of power; this reduces the need of DC/DC converters and channel separation but means as well that over voltages need to be switched off when they occur. Also it cannot be guaranteed that the MCUs won't be supplied by an overvoltage. Due to this, it is not possible to rely only on the switch off triggered by the MCUs but there is the need for an additional independent switch off (Discrete circuit). This architecture is shown as Figure 2

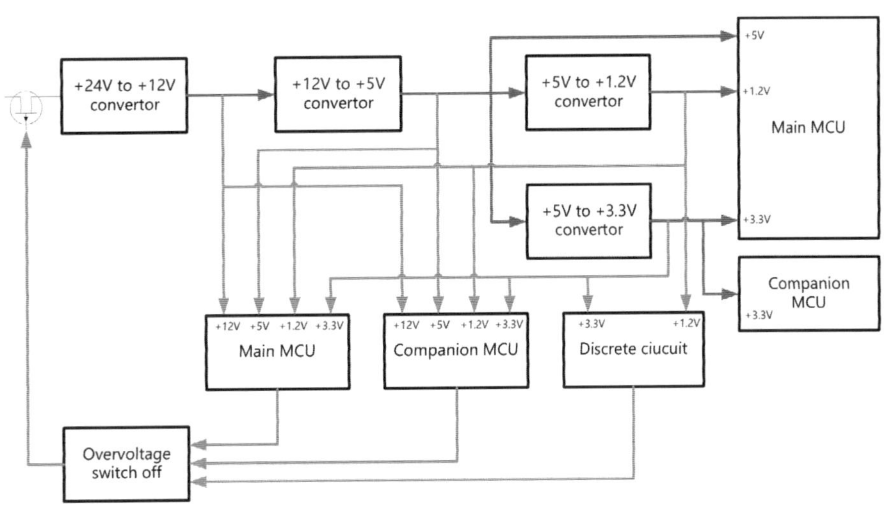

Figure 2. Power architecture

IOP Conf. Series: Materials Science and Engineering **922** (2020) 012003 doi:10.1088/1757-899X/922/1/012003

Both digital inputs and analogue inputs are implemented dual channel with just a single user terminal as Figure 3. This implementation is using redundancy instead of dynamic testing. Testing will be performed on status change of the input. The inputs are internally made as a 1oo2 system. The testing is done by comparison of the values that are read by the individual processors.
MOSFET are used to generate digital or PWM signals, for each channel there are 2 MOSFET, one controlled by main MCU and the other by companion MCU. Generally main MCU realize the normal function and companion MCU has the ability to shut off the channel when detecting failures.
The PWM current feedback is planned for non-safe control of the connected actuator and this signal can be used as diagnostic in closed loop current control mode. The current sense IC is dual channel and is therefore used to monitor two PWM channels. If the safety application has an expectation on the value the signal can be used for safety related purposes on application level.

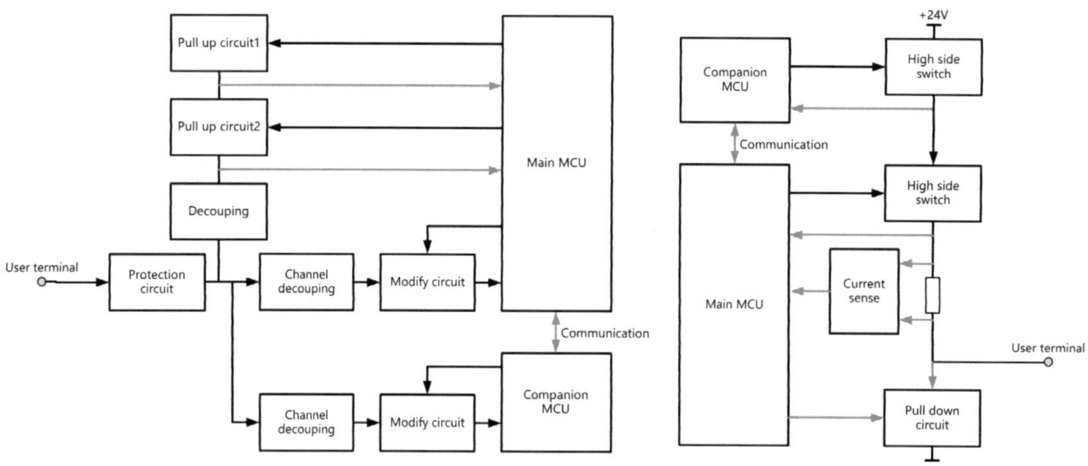

Figure 3. Input and Output design

3. Software design

Figure 4 shows the overview of the software architecture. The main MCU will run two parts of software, one is a CODESYS SIL2 runtime system and the other is the "Firmware" which contains all drivers for the peripheral as well as all diagnosis and self-test functions. Both software parts have to be able to exchange data with each other, for example the status of I/Os. All digital safe I/Os of the controller and one path of analogue inputs are located in the main MCU. The companion MCU is used for diagnosis and as second path for safe analogue inputs and as second switch-off path for the safe outputs. The data exchange between both MCUs occurs over the interface that is based on two SPI peripherals and some separate I/Os. The data include synchronous data, error signal, reset signal, watchdog signal and CRC check information.
Each of the both MCUs has two CAN-Interfaces. The main MCU uses these interfaces for communication with external sensors and for installation of software updates. The CODESYS SIL2 contains a CANopen safety stack for this. The companion MCU uses this interface only for future growth reasons. The companion MCU CAN Interfaces are only planned for non-safe data exchange.
There are several operating modes of the whole system as Figure 5. These modes are used for both MCUs but each mode can be implemented differently in each MCU because of different tasks. They are Fork-Fixed, Fork-Mode, Startup-Test-Mode.
The Fork-Fixed is executed first of all after the start of the system. This part is defined to make it possible to call the bootloader in case of an update error of the Fork-Mode. If the Fork-Mode would be called directly after the system startup and a software update will fail, the bootloader and firmware of the system cannot be reached any more.
The Fork-Mode will be called either directly after startup or by the Fork-Fixed part. This mode will decide whether the system starts the bootloader or whether the normal application software will be started.

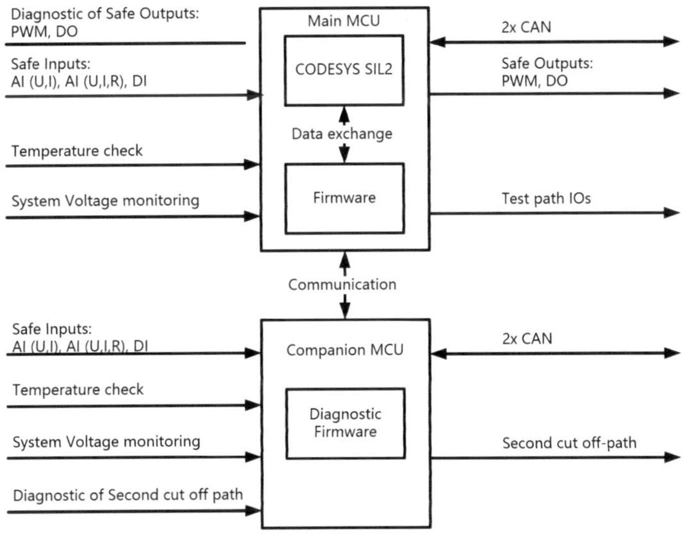

Figure 4. Software architecture overview

In Startup-Test-Mode, if an error is detected during one of the tests, the system change from "Startup-Test-Mode" into the "Safe-Mode". Otherwise, if all tests passed, the system will switch to "Init-Mode".

In Init-Mode, all peripherals of the MCUs are initialized first. Then the Retain-Data are loaded and checked for validity. The Retain-Data are data, which are provided by the CODESYS SIL2 Runtime. During the Startup this data must be read back. Afterwards the Init-Mode checks the software version of both MCUs for compatibility. If no error occurs, the system changes from Init-Mode into the Sync-Mode, otherwise the system will change to Safe-Mode.

The Sync-Mode is used for synchronization of both MCUs and could only be reached if the initialization in the Init-Mode has been successfully completed and all startup tests finished without an error. Both MCUs stay in this state until both could change at the same time into the Operating-Mode or a timeout. In case of a timeout both switch over into the Safe-Mode.

Operating-Mode is the normal state, during the Operating-Mode, both MCUs run different schedulers according to practical application.

The Safe-Mode is always activated in case of an error which affects other system functions as well. Both MCUs could change independent into the Safe-Mode and must then also transfer the other MCU to the Safe-Mode.

The Bootloader-Mode can be reached only from the Fork-Mode. If this Mode is entered, a non-safety relevant Bootloader application is executed. The only exit of the Bootloader-Mode is through a Power-Cycle.

The Debug-Mode is only reached from the Operating-Mode, after the CODESYS Runtime has been started. The Debug-Mode can be activated only by the CODESYS Runtime system. In this Debug-Mode the same functions are implemented as in the Operating-Mode. The difference between these two Modes is that the watchdog and all time relevant tasks are deactivated. The Debug-Mode is used for debugging the installed CODESYS Application.

4. Conclusion

This article describes the development of one high-performance safety controller for excavators; mainly include the hardware architecture and software architecture. The design focus is not only on function realization, but also on functional safety as the primary goal of design. Compared with standard controllers, this controller will has better performance and higher reliability in harsh environment. The design of hardware and software architecture of this controller is not only suited for excavators, but also other construction machineries.

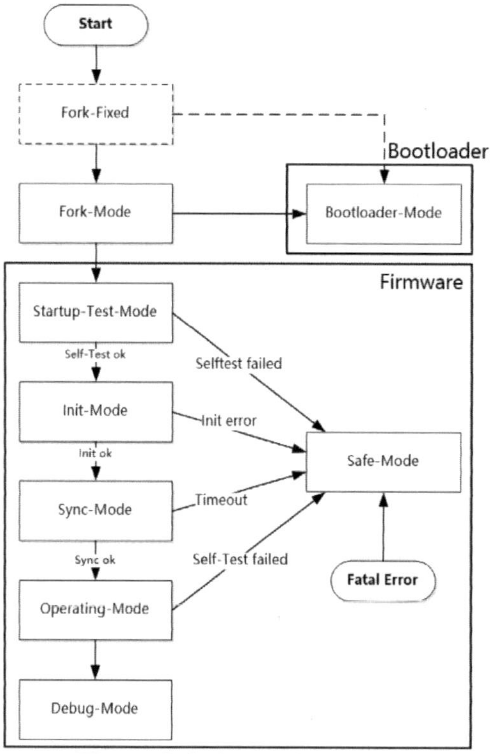

Figure 5. Software float overview

5. References

[1] IEC 61508-2010: Functional safety of electrical /electronic /programmable electronic safety-related systems.
[2] BS EN ISO 13849:2006: Safety of machinery Safety-related parts of control systems.
[3] Buja,Q Menis,R, 2012 *Dependability and Functional Safety: Applications in Industrial Electronics Systems*, Industrial Electronics Magazine, IEEE, 2012, 6(3): 4-12
[4] TMS570LS3137 16- and 32-Bit RISC Flash Microcontroller datasheet (Rev. C)
[5] Beckschulze, Eva,Salewski, Falk,Kowalewski, Stefan, 2009 *A comparison of dual-core approaches for safety-critical automotive applications*,. SAE International Journal of Passenger Cars - Electronic and Electrical Systems

IOP Conf. Series: Materials Science and Engineering **922** (2020) 012004 doi:10.1088/1757-899X/922/1/012004

Energy harvesting from pneumatic secondary suspensions for vehicle monitoring

A Genovese[1], G A D'Angelo[1], S Strano[1] and M Terzo[1]
[1] Department of Industrial Engineering University of Naples Federico II Naples, Italy

Abstract. In this paper, an activity concerning the design and the development of a vibration-based energy harvesting system is described. It consists of a piezoelectric based resonant device integrated in an air spring. A simulation study, together with the first experimental tests, have been carried out. The activity shows the goodness of the proposed solution as new concept of self-powered monitoring systems.

1. Introduction
Modern railway vehicles provide sophisticated monitoring systems for data acquisition and processing both for rolling stock and infrastructure condition; they must be fitted with high-capacity communication buses and multiple sensors in order to allow advance processing for data collection and management.

From a practical point of view, more sensors allow monitoring and diagnostics of the system, but it means, also, more cables to feed the sensors and to carry the signals from the field to the control room. The presence of cables on the bogies is not a desirable condition because it often creates several operative problems as installation, disconnection, long time to install cables and bogie maintenance.

Sensors powered by energy harvesting (EH) generators could be a suitable solution to overcome the cables issue and to face the environmental issues concerning the use of the batteries. Indeed, energy harvesting generators are become very attractive as inexhaustible replacements for batteries in low-power wireless electronic devices and have growing as a prominent research area in recent years.

Researchers have investigated many technologies to convert solar, heat, and vibration energy into electricity by energy harvesting systems. The main methods for energy harvesting on board system are electromagnetic [1] and piezoelectric [2][3].

This paper presents the development of an energy harvester integrated in a secondary suspension pneumatic spring. The system aims to recovery the vibration energy of the train suspensions by means of piezoelectric transducers in order to power a sensor and to give useful information for the monitoring and the diagnostics of vehicle or its subsystems [4] [5].

This document is organized as follows. Section 2 introduces the methodology adopted to carry out the harvester final project. Starting from the description of the chosen pneumatic spring, the main geometric characteristics and the expected design requirements are described. Then the harvester finite element design will be shown. Section 3 summarizes the obtained results and the designed EH solution, integrated in the pneumatic spring. Section 4 provides the concluding remarks.

2. Energy harvesting in the air spring
A commercial pneumatic spring has been chosen (figure 1) as a reference to perform the quantitative analysis in terms of energy harvesting, in both simulation and experimental environment. a reverse engineering process has been adopted to determine the geometric boundaries and the available volume reported in figure 2. The methodology adopted to carry out the final design is summarised in figure 3.

IOP Conf. Series: Materials Science and Engineering **922** (2020) 012004 doi:10.1088/1757-899X/922/1/012004

It is focused on the aim of integrating the energy harvester in a pneumatic spring [6] [7] to exploit the existing and protected space available in the bellow.

Figure 1. Reference Air Spring

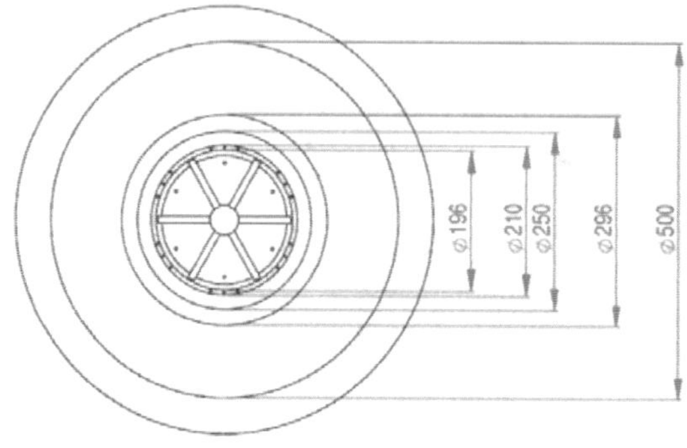

Figure 2. A 2D CAD representation of reference air spring with dimensions.

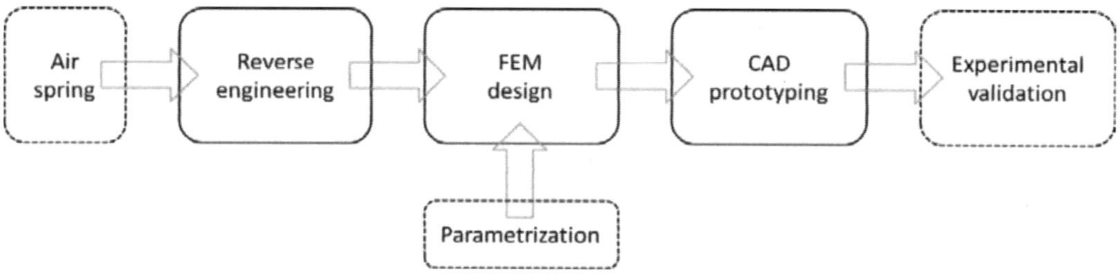

Figure 3. Schematich flow of adopted design procedure.

The reverse engineering process allowed to define the project requirements in order to create a faithful 3D model:

- The device must be integrated in the air spring
- The dimension of the device must respect a safe distance (50 mm) from the edge of air spring
- The device must be integral to the inlet plate of the air spring in order to exploit the excitations at the bogie frame that can range from 5 to 10 Hz

The idea is to collect energy through a resonant system and an ad hoc transduction mechanism using piezoelectric elements [8][9], not affecting the air spring functionalities. A preliminary test has been conducted on the shaker (figure 4) varying the amplitude and frequency of the base motion and the dimensions of the beam etc. These tests have been functional to characterize the main mechanical properties like stiffness and damping of the piezoelectric element (PZT255 d31operating mode). The main properties are illustrated in table 1.

Figure 4. Experimental layout for determination of piezoelectric beam parameters.

IOP Conf. Series: Materials Science and Engineering **922** (2020) 012004 doi:10.1088/1757-899X/922/1/012004

Table 1. Main parameters of piezoelectric patch.

Parameter	Description	Value	Unit
ρ	Density	7.80E+03	Kg/m^3
d_{31}	Piezoelectric charge coefficient	-1.74E-10	m/V
S_{11}^{E}	Elastic compliance	1.590E-11	m^2/N
$\varepsilon_{33}^{\sigma}$	Relative dielectric permittivity	1750	-
K_{31}	Coupling factor for transverse oscillation	0.351	-

A multi-physics finite element model of uniform cantilever beam has been developed in finite element environment [10][11]. The focus is on the electromechanical response. The cantilever beam layout, shown in figure 5, consists in a single layer of piezoelectric patch glued/bonded on a steel substrate with a tuning mass on the end.

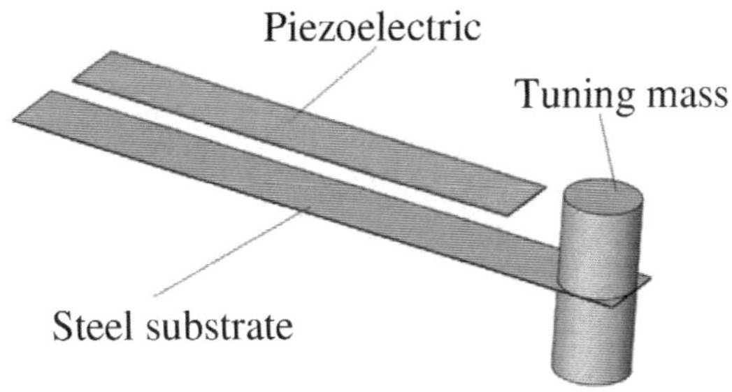

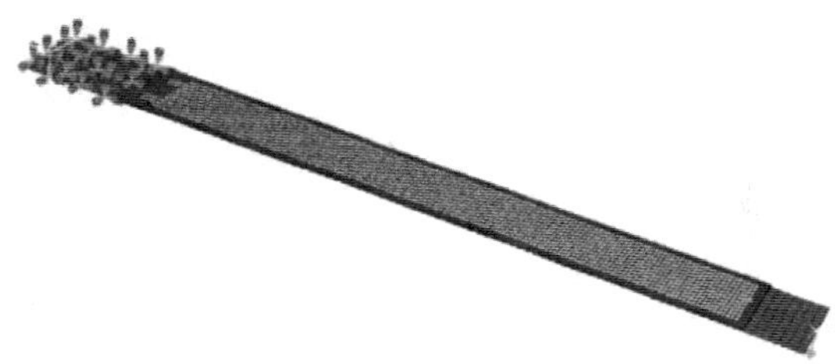

Figure 5. Cantilever beam layout a) fem model b).

The fem model is composed by two parts: steel substrate, modelled by shell elements, and piezoelectric patch. This element is composed by an elastic substructure, also called passive layer, to which the flexible piezoelectric films are attached. A perfect coupling between the substrate and the piezoelectric is defined with a surface-based tie constraint. In order to find the best geometry of the EH systems, in accordance with the design constraints, a parametric study has been conducted to evaluate the performance of PZT bonded steel beam and to choose the appropriate layout., The geometrical characteristics of PZT is varied [12] to study the effect of the dimensions on the behaviour of the composite beam The final geometrical characteristics of the PZT-bonded beam are reported in table 2.

The FE analysis is performed in two steps: in the first stage, a static analysis is firstly performed for characterize the mechanical properties. Afterwards, a modal analysis is carried out to calculate the natural vibration frequencies. With a mass of 30g, the device natural frequency is set to a value of 6 Hz, compatible with the excitation range (5-10 Hz) deriving from the bogie frame during train motion. Notice that the excitation induced by track irregularity is a random process, with statistical content, which generates a broadband excitation. For this reason, the system response is more dependent on the vehicle dynamics characteristics than the particular irregularity profile. In the second stage, the piezoelectric effect is implemented to characterize the electrical behaviour, then a dynamic analysis is performed to calculate the generated power. Therefore, a sinusoidal base motion is assigned with constant amplitude and frequency equal to the first natural frequency of the device. To emulate the electrodes of the piezoelectric patch and to simulate a piezoelectric power extraction circuit, the voltages on the top and bottom surfaces have been coupled to the electrical potentials of the master nodes assigned to each surface using linear constraint equations. The electrical potentials can be monitored at these master nodes. The piezoelectric surfaces bonded to the substrate are assigned a zero electrical potential throughout the analysis [13]. In this way it is possible to simulate an open circuit voltage response.

Table 2. Main parameters of PZT- bonded beam

Parameter	Steel substrate	Piezoelectric patch	Unit
Length	95	81.5	mm
Width	13	13	mm
Thickness	0.2	0.5	mm

3. Result.

Simulations have been performed for typical excitations in the railway vehicle in order to analyze the multi-physical behavior of the energy harvesting system and to evaluate the voltage generation to properly power a sensor. The results of the simulation have been used to build the physical prototype. Moreover, experimental tests have been conducted on the energy harvesting system highlighting the functionality of the proposed solution for sensors power supply and validating the design procedure. A full-scale mock-up has been developed by means of 3D printing (figure 6) to evaluate the dynamics of the system on the shaker, under the real working condition.

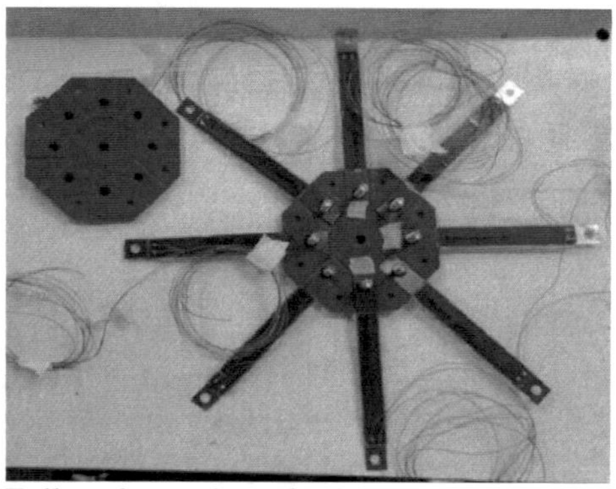

Figure 6. Full-scale mock-up for laboratory testing activities.

The prototype, composed by an octagonal support with eight piezoelectric beams, has been tested on the shaker. The tests have been conducted applying sinusoidal input with constant amplitude (2.2 mm) and frequency equal to the first natural frequency of the device. Figure 7 shows the comparison, in terms of frequency, of the voltage generated by the fem analysis and experimental tests. Figure 8 illustrates the comparison between input and output displacement, obtained in simulation (dashed line) and by experimental test, for a 6 Hz base excitation. It is possible to note that the simulation results are in good agreement with the experimental test.

Based on the geometric properties of the air spring and the requirements of the project, the selected solution for integrating the harvester into the air spring is shown in figure 9. The designed system consists of a steel support beam clamped between two elements that allow fixing the device to the air spring inlet plate. For this purpose, a fixing system has been developed, consisting of three steel elements. The lower disk plate (A) enables to clamp the system on the air spring and to fix the cantilevers beam support (B). The support of the cantilever beam presents a groove to host the beam, avoiding rotation around the vertical axis and translation along the transversal one. An upper plate (D), positioned on the support (B), completes the system ensuring the clamping, and locking the remaining degree of freedom. Figure 10 illustrates the EH solution embedded in the air-spring.

IOP Conf. Series: Materials Science and Engineering **922** (2020) 012004 doi:10.1088/1757-899X/922/1/012004

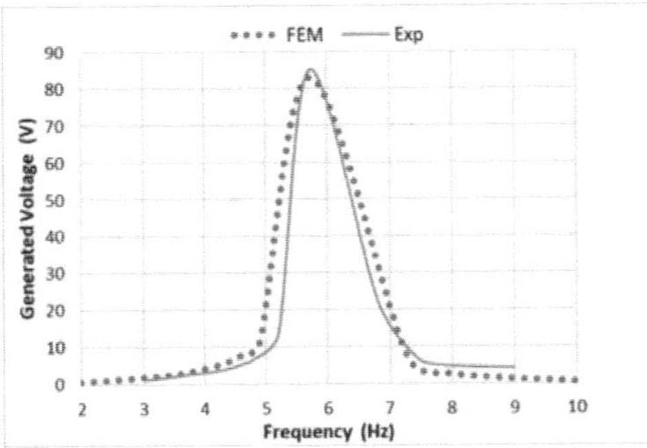

Figure 7. Comparison between the voltage generated by finite element analysis (dashed line) and experimental test.

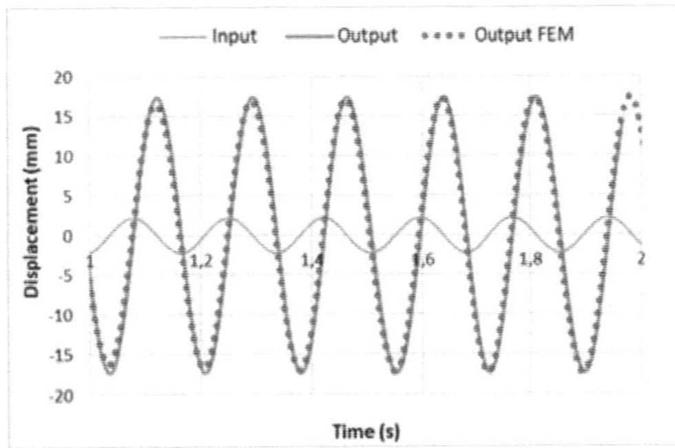

Figure 8. Comparison between input and output displacement, simulation and experimental test (dashed line).

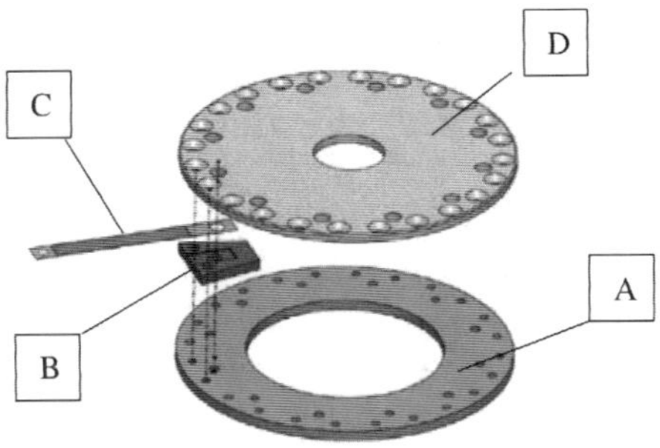

Figure 9. Final layout of EH integration system, lower disk plate (A), cantilever beam support (B), piezoelectric beam (C) and upper plate (D).

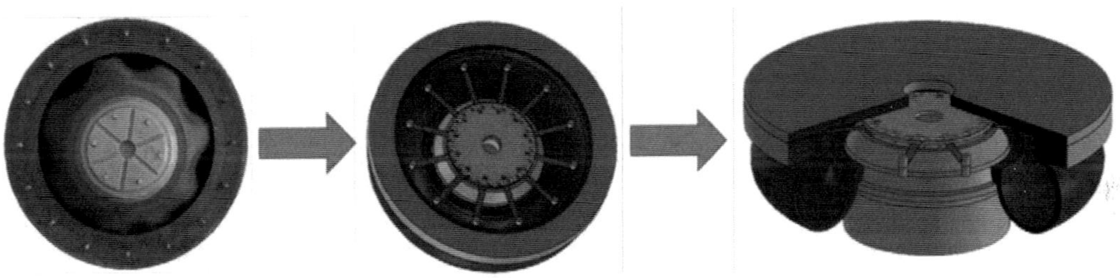

Figure 10. EH system embedding process.

4. Discussion and conclusions

This paper presents the prototyping of a motion-based piezoelectric energy harvester device, embedded in the secondary suspension of a high-speed train. The solution sims to overcome the cables issue and environmental concerns related to the use of the batteries.

The challenge has been to develop an advanced energy harvester that could be integrated into the pneumatic spring without requiring any significant modification of the component itself and, obviously, without interfering with its operating mode. On this way the existing air springs could be easily converted in smart components. The aim of the smart component is to convert mechanical energy of the bogie vibration, generated during the train ride, in electrical energy usable to power sensors. For this purpose, a specific design procedure based on the reverse engineering technique and on the finite element method has been carried out. The comparison of the FE model predictions against the experimental results shows a very good agreement both in terms of mechanical and piezoelectric behavior. The maximum output voltage is achieved near the frequency of 6Hz to exploit the excitation range deriving from the bogie frame during train motion.

Future developments should focus on the implementation of an electrical circuit to manage and to optimize the electricity generated in order to use it to power sensors useful for providing information for monitoring and diagnostics of the vehicle or its subsystems. In terms of hardware, further investigation could be conducted to test the device durability. Indeed, the system is designed to work in resonant condition and this situation could be a cause for concern. Anyway, in the case of the prototype device, no problems occurred after numerous experimental tests

5. References

[1] Y. Pan, F. Liu, R. Jiang, Z. Tu, and L. Zuo, "Modeling and onboard test of an electromagnetic energy harvester for railway cars," *Appl. Energy*, vol. 250, pp. 568–581, Sep. 2019, doi: 10.1016/j.apenergy.2019.04.182.

[2] G. De Pasquale, A. Somà, and F. Fraccarollo, "Piezoelectric energy harvesting for autonomous sensors network on safety-improved railway vehicles," *Proc. Inst. Mech. Eng. Part C J. Mech. Eng. Sci.*, vol. 226, no. 4, pp. 1107–1117, Apr. 2012, doi: 10.1177/0954406211418158.

[3] J. Y. Cho *et al.*, "Piezoelectric energy harvesting system with magnetic pendulum movement for self-powered safety sensor of trains," *Sensors Actuators, A Phys.*, vol. 250, pp. 210–218, Oct. 2016, doi: 10.1016/j.sna.2016.09.034.

[4] A. Genovese, S. Strano, M. T.-2019 I. 5th International, and undefined 2019, "Study of a Vibration-Based Piezoelectric Energy Harvester Embedded in an Air Spring," *ieeexplore.ieee.org*.

[5] F. D'Aniello, M. Riccio, … L. M.-2019 I. 5th, and undefined 2019, "Development and optimization of an energy harvesting circuit for multiple piezoelectric elements integrated into a smart air spring," *ieeexplore.ieee.org*.

[6] S. Strano and M. Terzo, "Molla ad aria con sistema di recupero energetico," 102016000128745 (UA2016A009238).

[7] A. Genovese, S. Strano, and M. Terzo, "Design and multi-physics optimization of an energy harvesting system integrated in a pneumatic suspension," *Mechatronics*, vol. 69, p. 102395, Aug. 2020, doi: 10.1016/J.MECHATRONICS.2020.102395.

[8] N. G. Stephen, "On energy harvesting from ambient vibration," *J. Sound Vib.*, vol. 293, no. 1–2, pp. 409–425, May 2006, doi: 10.1016/j.jsv.2005.10.003.

[9] Genovese A, Strano S, and Terzo M, "Model-based study of a resonant system in a railway air spring secondary suspension for energy harvesting applications," *dl.acm.org*, vol. 2018-February, pp. 45–51, Feb. 2018, doi: 10.1145/3185066.3185073.

[10] C. Junior, A. Erturk, D. I.-J. of S. and Vibration, and undefined 2009, "An electromechanical finite element model for piezoelectric energy harvester plates," *Elsevier*.

[11] W. S. Hwang and H. C. Park, "Finite element modeling of piezoelectric sensors and actuators," *AIAA J.*, vol. 31, no. 5, pp. 930–937, 1993, doi: 10.2514/3.11707.

[12] M. Aslam, P. Nagarajan, and M. Remanan, "Numerical Studies on PZT-Bonded Aluminium Beam," *J. Inst. Eng. Ser. A*, vol. 100, no. 1, pp. 117–130, Mar. 2019, doi: 10.1007/s40030-018-0340-5.

[13] A. Abdelkefi, N. Barsallo, L. Tang, Y. Yang, and M. R. Hajj, "Modeling, validation, and performance of low-frequency piezoelectric energy harvesters," *J. Intell. Mater. Syst. Struct.*, vol. 25, no. 12, pp. 1429–1444, 2014, doi: 10.1177/1045389X13507638.

Acknowledgments

The authors would like to thank Monticelli Group srl for the support given in the activity

IOP Conf. Series: Materials Science and Engineering **922** (2020) 012005 doi:10.1088/1757-899X/922/1/012005

A Physical-Based Observer for Vehicle State Estimation and Road Condition Monitoring

B Lenzo[1], G Ottomano[2], S Strano[2], M Terzo[2] and C Tordela[2]

[1]Sheffield Hallam University, United Kingdom
[2]University of Naples Federico II, Italy

Abstract. The performance of the vehicle's active safety systems depends on accurate knowledge of the vehicle state, and the frictional forces resulting from tyre contact and the road surface. This paper aims to estimate the vehicle states and tyre-road coefficient of friction through and Extended Kalman Filter (EKF), integrated with the Double-Track model and the Pacejka Magic Formula that allows knowledge of the lateral force of the tyre. Besides, this approach can estimate the overall coefficient of lateral friction on each side of the vehicle, left and right respectively. Simulations based on a reference vehicle model are performed on different road surfaces and driving manoeuvres to verify the effectiveness of the proposed estimation method, in order to obtain good performance from different vehicle control systems.

1. Introduction

Over the past decades driver assistance system have become a standard in automotive industry [1]. Nevertheless, the number of deaths caused each year in the world by road accidents exceeds one million. This number is unacceptable as technology and technology advances. In order to try to reduce this very high value, it is possible to try to improve the performance of these driver assistance systems. The performance of these systems could be further improved if more accurate knowledge on the vehicle state, inputs and parameters would be available. However, many of these variables, as the sideslip angle, cannot be measured directly in commercial vehicles because the sensors are very expensive. In literature, it's possible to distinguish mainly two approaches to develop vehicle state observers, that employing readily available sensors to correct the estimation of the variables which require the employment of expensive sensors. The first approach uses a kinematic vehicle model, independent from tyre parameters and road condition, in combination with measurement from standard vehicle sensor. This estimation technique is sensitive to sensor errors (noise and bias). A correction for these errors by GPS measurement is possible [2] [3], but require accuracy is not achievable by consumer-grade GPS and reception may be lost.

The second approach uses a dynamic vehicle model in combination with the measurements from standard vehicle sensors. With this approach the model can correct for sensors inaccuracies and unwanted measurements, but information on tyre parameter and road condition is needed for the tyre model. Many observers have been developed based on this approach [4] [5] [6] [7] [8], where different tyre models have been used.

This paper proposes a new method for estimating the fundamental variables of vehicle dynamics and, at the same time, thanks to the use of a simple Magic formula characterized by four parameters obtained from extensive offline testing [9][10], the estimation of the lateral friction coefficient, through the use of an extended Kalman filter. The determination of interaction forces is a very complex operation, and as already mentioned, it requires the adoption of very expensive sensors. Through the approach proposed in this work a computationally slim method has been offered,

IOP Conf. Series: Materials Science and Engineering **922** (2020) 012005 doi:10.1088/1757-899X/922/1/012005

one that can be inserted inside a car control unit and therefore considerably reduce the cost of sensors on the vehicle. And in the final analysis to be able to allow to equip, any type of car, with high-performance safety systems and driving aid.

2. Estimator Design

The estimator is based on the Double-Track model. This model is shown in Figure 1 and Figure 2. The nomenclature used follows the following standard: the first subscript indicates the axle (front/rear), while the second subscript indicates the position (left/right). δ is the steering wheel angle; r is the yaw rate; V_G is the centre of gravity (COG) velocity vector, v and u are, respectively , the COG vehicle velocity components in lateral and longitudinal direction; β is the COG sideslip angle; $\alpha_{11} = \delta - atan\left(\frac{v+ra}{u-r\frac{t}{2}}\right)$, $\alpha_{12} = \delta - atan\left(\frac{v+ra}{u+r\frac{t}{2}}\right)$, $\alpha_{21} = -atan\left(\frac{v-rb}{u-r\frac{t}{2}}\right)$ and $\alpha_{22} = -atan\left(\frac{v-rb}{u+r\frac{t}{2}}\right)$ are respectively the front left, the front right, the rear left and the rear right tyre slip angles; F_{xij} and F_{yij} are the longitudinal and lateral tyre-road interaction forces; a and b are the distances from the COG to the front and rear axle; s is the vehicle track width; m is the vehicle mass and J is the yaw inertia moment. $\dot{v}$ and $\dot{r}$ represent the derivatives of v and r with respect to time.

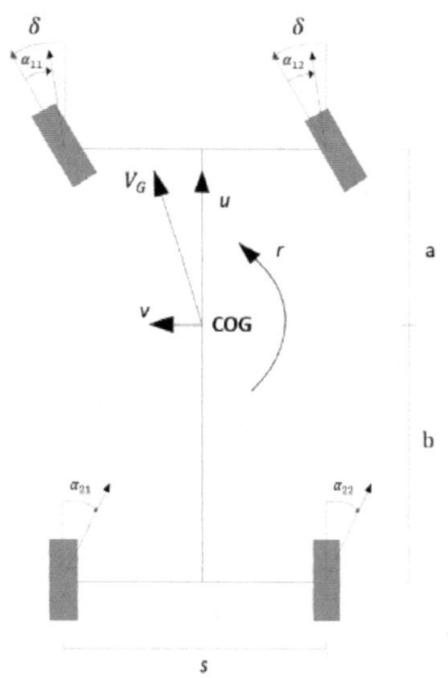

Figure 1. The Double-Track model: velocity vectors, slip angles and steering wheels angles.

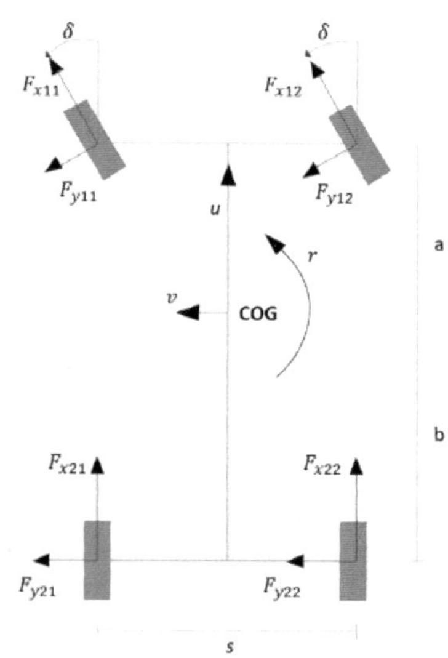

Figure 2. The Double-Track model: velocity vectors and pneumatic-road tangential interaction forces.

Table 1. List of vehicle main physical parameters.

Parameter name	Value
Vehicle mass	2217 kg
Distance from COG to front wheels	1.397 m
Distance from COG to rear wheels	1.663 m
Wheelbase	2.760 m
Height of COG	0.65 m
Yaw inertia moment	3231 kg/m^2
Track width	1.492 m

The lateral vehicle dynamic can be formulated as follow:

$$\begin{cases} \dot{v}_y = \dfrac{1}{m}\left(F_{y11}\cos\delta + F_{y12}\cos\delta + F_{y21} + F_{y22}\right) - ur \\ \dot{r} = \dfrac{1}{J}\left(F_{y11}a\cos\delta + F_{y12}\cos\delta\, a - F_{y21}b - F_{y22}b + F_{y11}\sin\delta\dfrac{s}{2} - F_{y12}\sin\delta\dfrac{s}{2}\right) \end{cases} \tag{1}$$

A four-parameter version of the Pacejka Magic Formula (stiffness factor B, form factor C, peak value D, curvature factor E) is used to model the tyre/road interaction forces to complete the dynamic vehicle description:

$$F_y(\alpha) = D\sin\{C\,\mathrm{atan}[B\alpha - E(B\alpha - \mathrm{atan}\,(B\alpha))]\} \tag{2}$$

The parameter D is placed equal to the product between the global lateral friction coefficient and the vertical load, respectively equal to $\mu_l F_z$ for the left side of the vehicle, and equal to $\mu_{lr} F_z$ for the right side, obtaining two different formulations of the Pacejka Magic Formula (MF). Considering the global lateral friction coefficients represented by the following equation

$$\dot{\mu}_l = 0;\ \dot{\mu}_r = 0 \tag{3}$$

as additional states of an augmented state vector [11], the estimator design model is obtained. The state vector is given by $\boldsymbol{x} = [v,\ \ r,\ \ \mu_l,\ \mu_r]^T$ where v is the lateral velocity, r is the yaw rate, and μ_l and μ_r are the lateral coefficients of adhesion left and right respectively. The input vector $\boldsymbol{u} = [\delta,\ \ u]^T$ is composed by the steering wheel angle δ and the longitudinal speed of the vehicle u. The EKF estimation algorithm is briefly recalled in Figure 3 [11]. Consider the nonlinear state function $\boldsymbol{f}$ given by the combination of the equations (1) – (3) with an additive Gaussian process noise $\boldsymbol{w}$ and a nonlinear function $\boldsymbol{h}$ of measurement equations with an additive Gaussian noise $\boldsymbol{v}$:

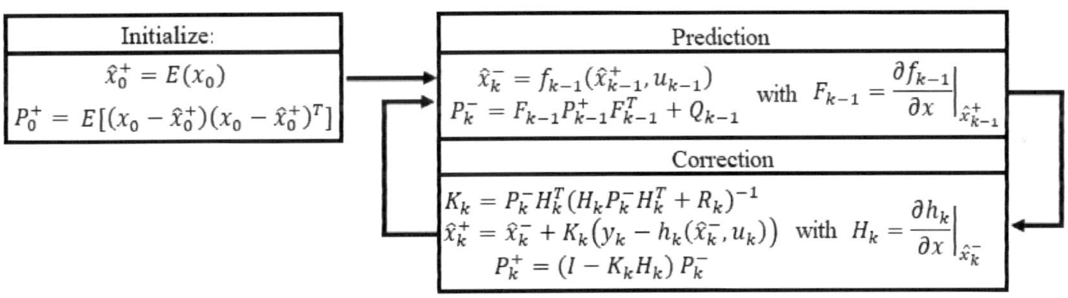

Figure 3. EKF estimation algorithm

Diagonal and constant process noise covariance Q_{k-1} and measurement noise covariance R_k matrices are considered. In the estimator design model outlined above, the estimated measurements vector is constituted by the lateral acceleration and the yaw rate:

$$\hat{y} = \begin{bmatrix} \hat{a}_y \\ \hat{r} \end{bmatrix} = \begin{bmatrix} \dot{v} + ur \\ r \end{bmatrix} \tag{4}$$

This vector is compared with the real measurements vector coming from the vehicle to perform the updating operation of the estimated states and the error covariance. The suitability of the EKF estimator for the proposed application is demonstrated for two case studies.

3. Simulation Results

The simulations results presented in this section are obtained using Matlab/Simulink. A comparison was made between the estimator and a full-vehicle model obtained through the software Adams/Car.

The sideslip angle, the left and right coefficient of frictions and the measurements of the reference vehicle are compared with the estimation provided by the EKF. In order to obtain realistic measurements, Gaussian noises are added in the simulated measurements. The simulation results of two representative manoeuvres are presented here. Table 2 gives the details and purpose of each manoeuvres. Figure 4 provides the indicative physical representation of each proposed scenario where the colour represents the friction coefficient. The black colour corresponds to a high friction value; the green colour corresponds to a medium friction value; and the green colour corresponds to a low friction value. The change in friction is time-dependent and occurs at successive time intervals. During simulations, the system is set on an equilibrium point with constant longitudinal velocity

Table 2. Details and investigative purpose of each simulated manoeuvres.

Simulated manoeuvres	Test Surface	Purpose
Step steer – case 1: step varying μ for all tyres	Five levels of global lateral friction coefficient with unequal step sizes	Determinate the estimator response to a quick change in road surface. Friction coefficient is varied at five new levels.
Step steer – case 2: different step varying μ for left/right tyres	Each wheel is subjected simultaneously to two different frictions	Validate the TRFC detection under small friction variations. Also, each wheel is exposed to different surface at the same time (distinct μ for the left and right tyres)

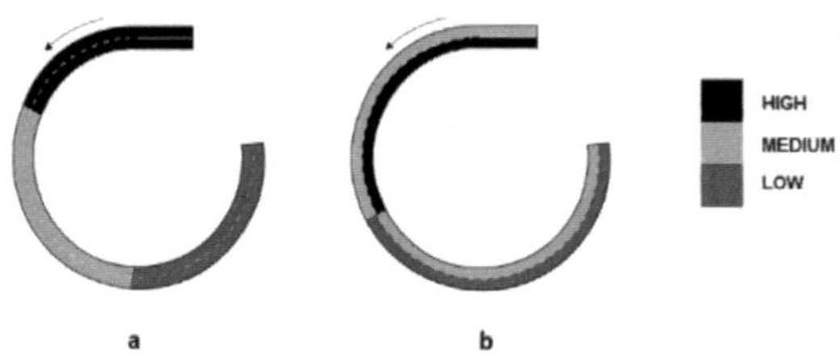

Figure 4. Schematic road layout with time-dependent changing of global friction coefficient.

3.1. Step Steer Manoeuvre: case 1

In this configuration, Figure 4a, a decreasing trend of the tire-road friction coefficient has been imposed. Transitions occur during successive equal time intervals of 10 seconds A $0 - 100\ deg$ change in the steering angle of the front left and right tires have been imposed. The longitudinal speed is set to $60\ km/h$. The estimator behaviour is represented by the measurements vector and the sideslip angle, Figure 5; the global left and right lateral friction coefficient, Figure 6; the lateral forces, Figure 7, and the normal, Figure 8, forces of the tire.

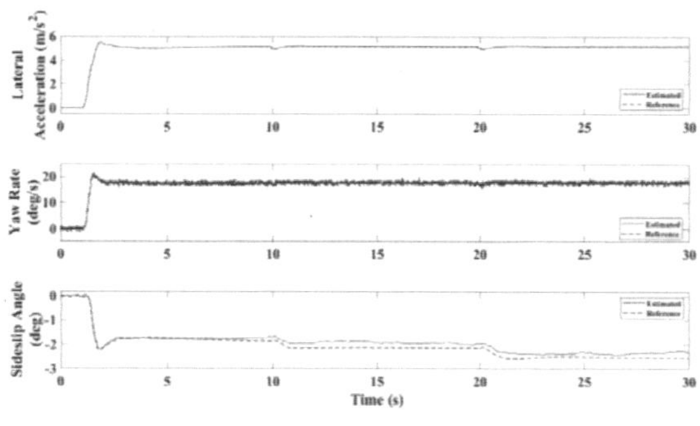

Figure 5. Measurements Vector and Sideslip Angle: Step steer case 1.

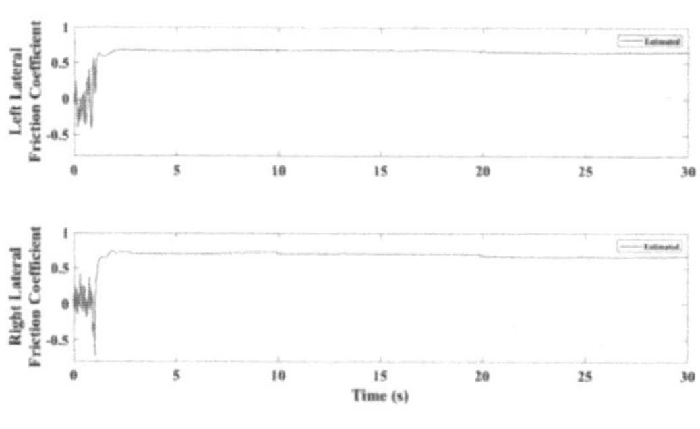

Figure 6. Left and Right lateral global friction coefficient in a time varying friction scenario.

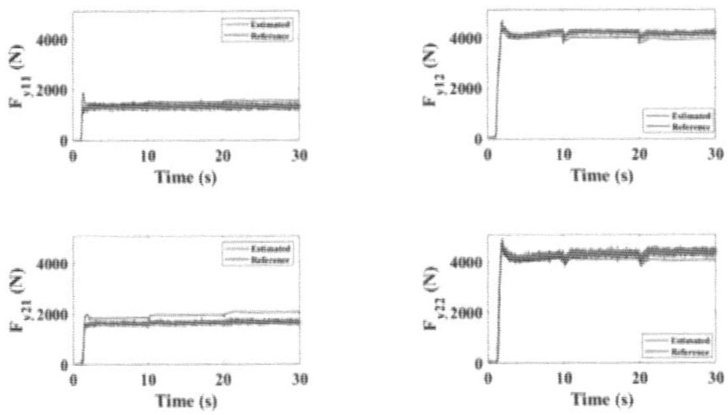

Figure 7. Lateral forces of the tyre: Step steer case 1.

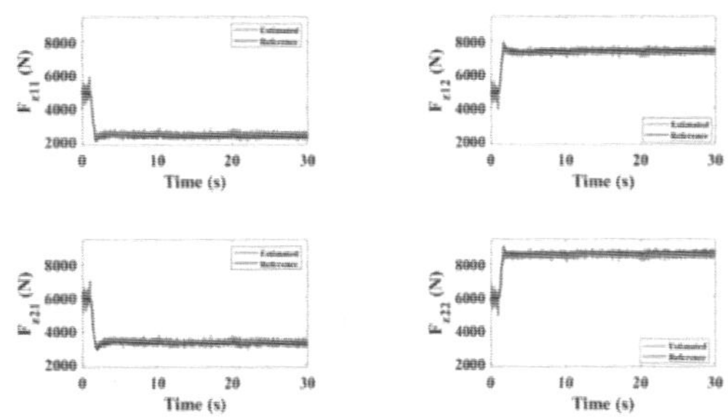

Figure 8. Normal forces of the tyre: Step steer case 1.

This simulation allows appreciating how the estimator reacts to successive variations of the global lateral friction coefficient. It is possible to observe three different variations of this latter value that occur every 10 seconds. The estimator manages, in an excellent way, to provide an estimation of the value of the global lateral friction coefficient and the vehicle states.

3.2. Step Steer Manoeuvre: case 2

While case 1 showed an equal change for all the wheels, here, the tyre-road friction coefficient is set at two different levels for the left and the right side of the vehicle, Figure 4b. The transition occurs during an equal time interval of 10 seconds. A $0 - 100\ deg$ change in the steering angle of the front left and right tires have been imposed. The longitudinal speed is set to $60\ km/h$. The estimator behaviour is represented by the measurements vector and the sideslip angle, Figure 9; the global left and right lateral friction coefficient, Figure 10; the lateral forces, Figure 11, and the normal, Figure 12, forces of the tire.

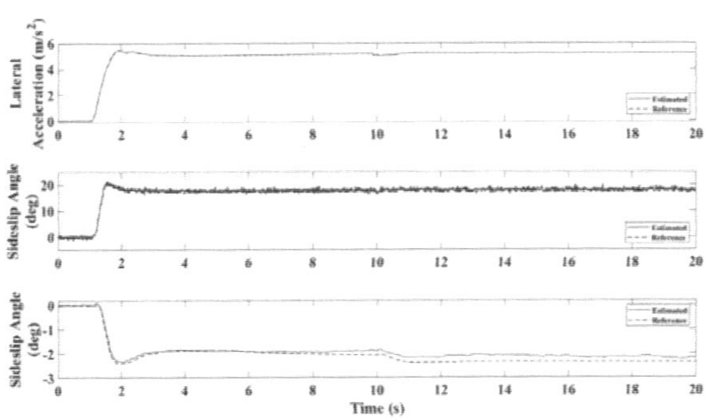

Figure 9. Measurements Vector and Sideslip Angle: Step steer case 2.

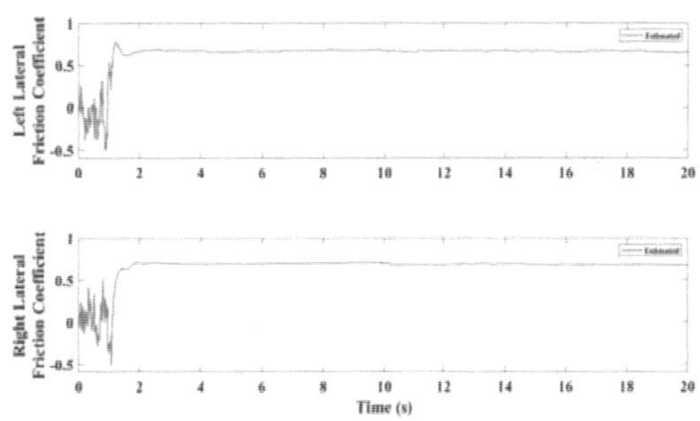

Figure 10. Left and Right lateral global friction coefficient in a time varying friction scenario with different friction for each side of the vehicle.

IOP Conf. Series: Materials Science and Engineering **922** (2020) 012005 doi:10.1088/1757-899X/922/1/012005

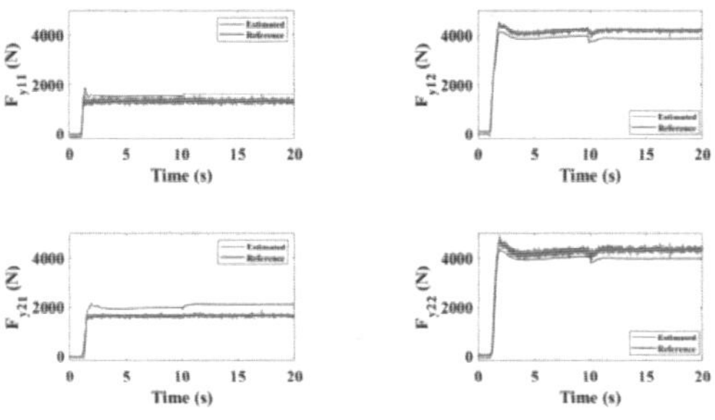

Figure 11. Lateral forces of the tyre: Step steer case 2.

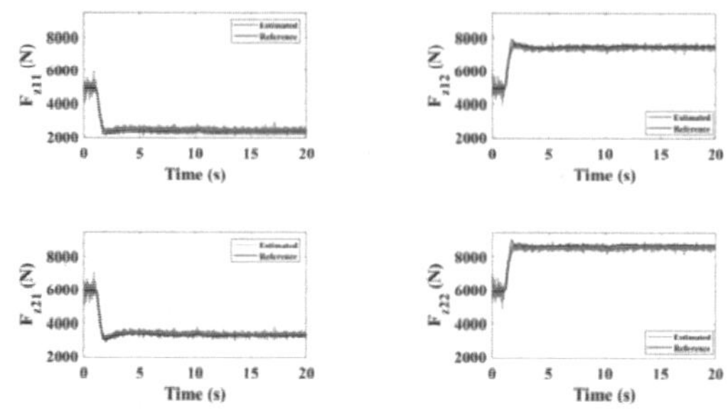

Figure 12. Normal forces of the tyre: Step steer case 2.

This simulation highlights a different peculiarity of the EKF. The estimator is able to detect a different value of the global lateral coefficient of friction on the two different sides of the vehicle and manages to detect the small variation in friction that occurs at 10 seconds. In both situations, the employed approach provides an excellent estimation of the vehicle conditions.

4. Conclusions

An Extended Kalman Filter has been proposed for the estimation of the lateral friction coefficient and the state of the vehicle. The procedure provides a reliable estimation of the state of the system and a consistent identification of the lateral friction coefficient. The EKF has been compared to a full-vehicle Adams/Car model, these simulations results are the first step in the model validation. This estimation can be used to improve the performance of the vehicle's active safety systems, as well as to monitor real-time changes in road conditions. A possible future development is its implementation in an onboard real-time system.

5. References

[1] Bosch R 2014 Bosch Automotive Handbook 9th ed.

[2] Ryu J and Gerdes J C 2004 Integrating inertial sensors with gps for vehicle dynamics control Journal of Dynamic Systems, Measurement, and Control 243–254.

[3] Strano, S., Terzo, M. Constrained nonlinear filter for vehicle sideslip angle estimation with no a priori knowledge of tyre characteristics (2018) Control Engineering Practice, 71, pp. 10-17.

[4] Strano, S., Terzo, M. Vehicle sideslip angle estimation via a Riccati equation based nonlinear filter (2017) Meccanica, 52 (15), pp. 3513-3529.

[5] Farroni, F., Lamberti, R., Mancinelli, N., Timpone, F. TRIP-ID: A tool for a smart and interactive identification of Magic Formula tyre model parameters from experimental data acquired on track or test rig, Mechanical Systems and Signal Processing, 102, pp. 1-22.

[6] Farroni, F, T.R.I.C.K.-Tire/Road Interaction Characterization & Knowledge - A tool for the evaluation of tire and vehicle performances in outdoor test sessions. Mechanical Systems and Signal Processing, 72-73, pp. 808-831.

[7] Boulkroune B, v Aalst S, Rademakers E, Sannen K and Ghouti N E 2015 Vehicle sideslip angle estimation using nonlinear parameter-varying observers In 2015 Int. Automatic Control Conference 108–113.

[8] Farroni, F., Pasquino, N., Rocca, E., Timpone, F. A comparison among different methods to estimate vehicle sideslip angle. Lecture Notes in Engineering and Computer Science, 2018, pp. 1084-1090.

[9] Pacejka H B 2006 Tyre and Vehicle Dynamics, 2nd ed.

[10] Tass “Tyre testing,” https://www.tassinternational.com/tyre-testing.

[11] Simon D 2006 Optimal State Estimation: Kalman, H∞ and nonlinear approaches. J. Wiley and Sons, New York

Modelling and simulation of damage mechanisms in railway vehicle-track interaction

F Ciuffini[1], F Favo[1], S Giulianelli[1], A Miele[2], G R Sitongia[1], S Strano[2]

[1]RFI, Rome, Italy
[2]Department of Industrial Engineering University of Naples Federico II, Naples, Italy

Abstract. This paper focuses on the estimation of damage contribution that each type of railway vehicle produces on the line Pescara - Ancona. The results of the numerical simulations provides fundamental information to understand the railway vehicle properties mostly influencing track damage mechanisms. The results have clearly shown the influence of the wheelset vertical forces on the damage indicators. The analysis of the specific contribution on damage by a specific railway vehicle will be particularly important to modulate track access charges by vehicle type. Moreover, simulation results will allow to adjust the unit maintenance cost of the line on the class of train, whose volumes are known.

1. Introduction

The study on the damage dynamics of a railway infrastructure is one of the most interesting topics in the railway field in recent decades. The deterioration of the rails is mainly caused by the passage of the rolling stock and also, it depends on some characteristics of the vehicles such as the mass, the speed, the type of traction and the wheelset distance, and on some characteristics of the railway infrastructure, such as the curvature radius in the curve sections.

In general, rails are subject to damage due to the interaction forces exchanged by railway vehicles. This leads to maintenance and safety issues. The damage is often categorized as rolling contact fatigue (RCF), track settlement and wear of the rail [1]. This paper is focused on the estimation of damage contribution that rolling stocks produce on the line Pescara-Ancona. The plano-altimetric characteristics of this line are typical of a straight plain line, with few corners. It is a mixed traffic line with regional, long haul and freight trains: this line can be taken as the representative line of a large part of the italian network.

Vehicles contribute to track deterioration in different way, depending on the different interactions between vehicle and track. These damages are the causes of maintenance actions such as tamping, rail treatment and rail replacement. Engineering models are generally used to modulate costs between different types of traffic, while these costs are recorded with other methods (econometric or allocative). On a new infrastructure, only engineering models can be used. The proposed method has the aim of comparing some significant wear parameters regarding the track, modulating a measure of damage obtained with other models. Infrastructure elements subject to wear during the passage of trains are shown in Figure 1 and the ordinary maintenance of these items is due to direct costs. A large part of the maintenance of tracks, junction, ballast could be carried out by the Infrastructure Managers (IMs) as extraordinary maintenance in the context of line renewals and the related costs are not attributable as direct costs.

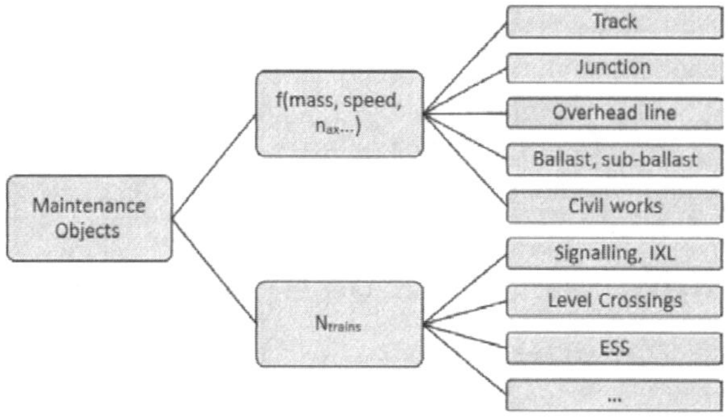

Figure 1. Maintenance objects

In this model, items are divided into 2 categories: items subject to variations in the physical features of the train, such as mass, speed and number of axles, and items variable only in relation to the number of trains passing through. Focusing on track damage, an engineering damage function is investigated, given track geometry and train features, finding most significant ones, and as an output a wear damage level index.

In order to achieve that, the software SIMPACK® Rail has been used for the simulations. SIMPACK® Rail is the world-wide market leader for rail vehicles multi-body simulation with a wide dataset of specific components for the construction of railway vehicle models and for the detailed modelling of the wheel-rail contact [2]. In addition, the SIMPACK® Rail Wear add-on module has been used for the damage calculation. Simulations with this tools provided comparison of vehicle concepts concerning wear and the automated calculation of damages.

2. Damage mechanisms

To estimate the damage functions it is necessary to calculate 3 quantities: dynamic forces for single wheel, wheel/rail contact area and the specific friction energy. The dynamic force for single wheel is:

$$Q(V) = a_Q \cdot V + b_Q \tag{1}$$

where Q(V) is the dynamic vertical force at the wheel, V is the vehicle speed, a_Q and b_Q are coefficients that depend on vehicle and rail characteristics.

The contact area A_{wheel} is calculated with dimensions of the two semi-axis of the ellipse *a* and *b*:

$$A_{wheel} = \pi \cdot a \cdot b \tag{2}$$

Once the dynamic forces and contact area have been calculated, these are the inputs for the different damage laws that allow to calculate the damage indices D_1, D_2, D_3 $D_{4.1}$ and $D_{4.2}$.

For the relationship between vertical dynamic force at wheel Q(V) and ballast damage / rail displacement it is assumed that $D_1 \sim Q^3$:

$$D_1 = n_{RS} \cdot Q(V)^3 \tag{3}$$

where n_{RS} is the number of wheelsets which contribute to damage. The maintenance activity associated with D_1 is tamping (levelling/straightening). The relationship between the dynamic vertical force at the wheel Q(V) and the rail defects on the straight track it is assumed that $D_2 \sim Q^{1,2}$:

$$D_2 = n_{RS} \cdot Q(V)^{1,2} \tag{4}$$

The Traction power value (Tpv) is considered as an index of rolling contact fatigue caused by traction of the driven wheel.

$$TPV = D_3 = \frac{P_{wheel}}{A_{wheel}} \tag{5}$$

The specific energy of friction W_b in the wheel-rail contact is strictly related to rolling contact fatigue $D_{4.1}$ and wear $D_{4.2}$ [3].
The functions of $D_{4.1}$ and $D_{4.2}$ are:

$$\begin{cases} D_{4.1} = 0 & for \quad W_b < 15\,N \; and \; W_b \geq 175\,N \\ D_{4.1} = n_{FW} \cdot (0.02 \cdot W_b - 0.3) & for \quad 15\,N \leq W_b < 65\,N \\ D_{4.1} = n_{FW} \cdot (-W_b + 175)/110 & for \quad 65\,N \leq W_b < 175\,N \end{cases} \tag{6}$$

$$\begin{cases} D_{4.2} = 0 & for \quad W_b < 65\,N \\ D_{4.2} = n_{FW} \cdot (W_b - 65)/110 & for \quad W_b \geq 65\,N \end{cases} \tag{7}$$

where n_{FW} is the number of leading wheelsets. As can be seen, for high values of W_b the phenomenon of rail wear increases while contact fatigue due to rolling disappears [3]. The maintenance activities associated with D_2, D_3 and $D_{4.1}$ are rail treatments like grinding or milling while rail replacement is related to $D_{4.2}$.
The specific friction energy W_b, also known as wear number, has been evaluated with MBS A multibody model of a passenger vehicle was generated starting from the wheel/rail contact up to the definition of the entire train, as can be seen in Figure 1 and Figure 2, specifying all interactions between the various components.

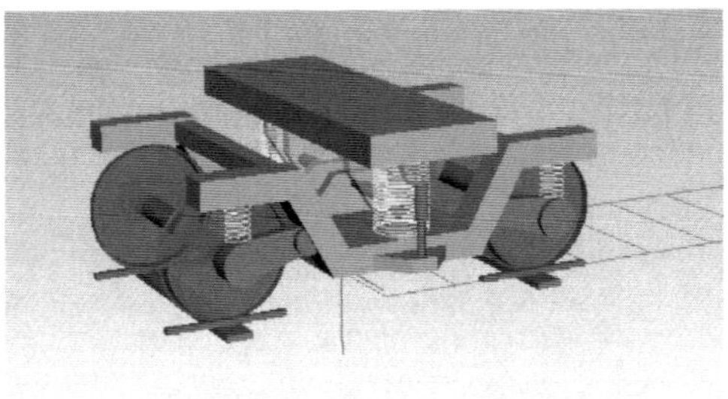

Figure 2. Bogie dynamic model

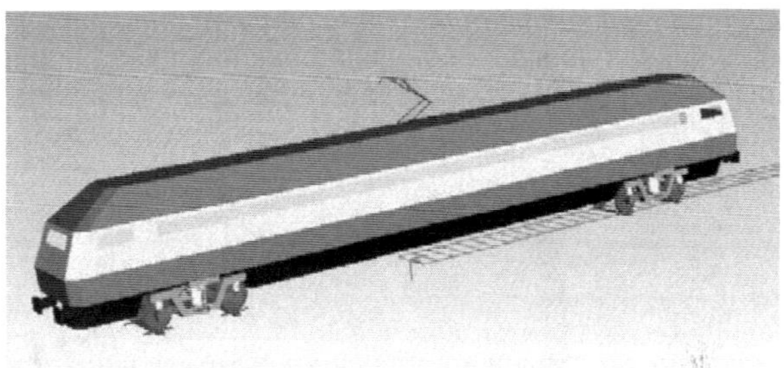

Figure 3. Train dynamic model

According to the dynamic simulations results, the trend of the wear number was determined in function of the variation of 4 parameters featuring rail infrastructure or rail vehicle. These parameters are curvature radius in the curve sections, mass of the vehicle, longitudinal stiffness of the primary suspension and the wheelset distance.

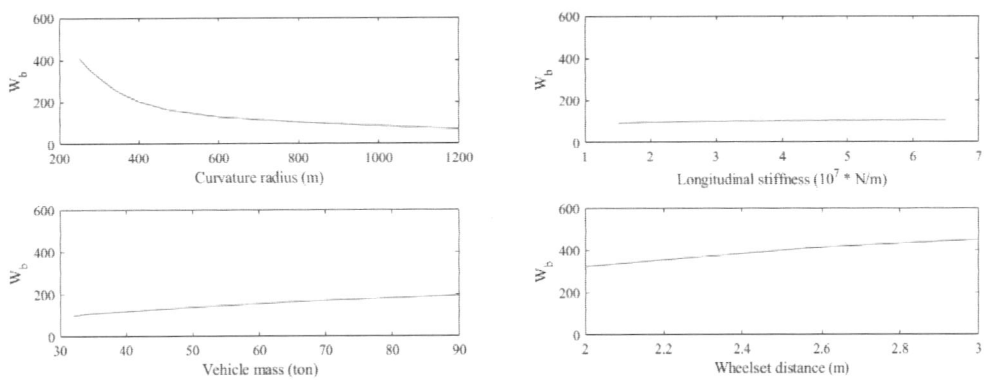

Figure 4. Wear number trend

As can be seen in Figure 4, the wear number has different trends depending on which parameter is being varied. Each time only one parameter has been changed while the others remain constant. As the curvature radius increases, the W_b decreases. The vehicle mass and the wheelset distance influence the W_b trend in the same way. As the longitudinal stiffness increase, the W_b trend remains almost constant.

3. Results

The test case track considered for the simulations is the Pescara-Ancona line. For this line track a dynamic model was generated in SIMPACK® considering the planoaltimetric profile, the rail inclination (1:40) and the cant in the curve sections. A model example can be seen in Figure 5.

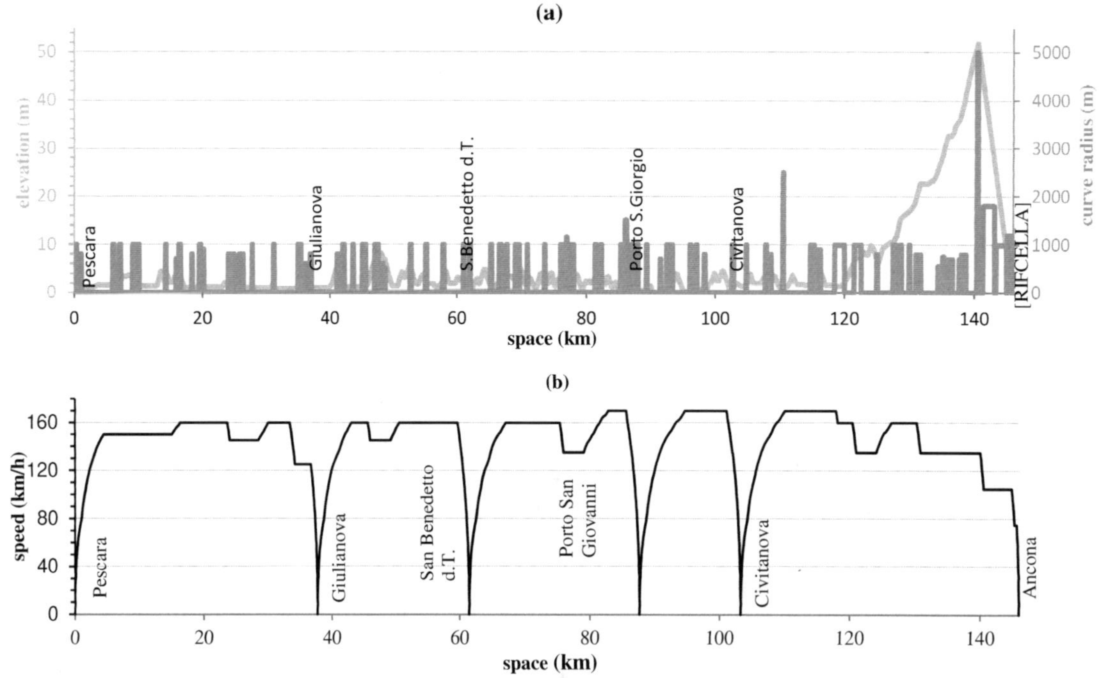

Figure 5. Track Pescara-Ancona: (a) planoaltimetric profile; (b) speed profile

In this paper, 8 different passenger rolling stocks are considered, which represent 4 categories of service performed.

Table 1. Characteristics of the rolling stocks

Rolling Stock	Locomotives + Coaches	Average speed (km/h)	Total mass (t)	Average axle-load (t)	Service performed
Ale 501	3	86	123	15	Regional fast
ETR 425	5	86	205	17	Regional fast
E 414 + UIC Z1	2 + 9	153	577	13	High speed
E 444 + UIC Z1	1 + 7	128	330	10	Intercity
E 464 + UIC Z1	1 + 5	68	322	13,5	Regional slow
ETR 324	4	86	164	16,4	Regional fast
ATR 220 tr	4	68	180	18	Regional slow
ETR 500	2 + 11	153	576	11	High speed

In Table 1 there are some characteristics of the trains considered in this paper. In the first column there is the composition of each train in terms of locomotives and coaches. In the second one there is the average speed calculated by speed profile of each train. The third one represents the mass of each train and the fourth column is the ratio between the mass of a train and the wheelsets number. The last column define the service performed by each vehicle.
In Figures 6 and 7, it is possible to observe the results in terms of damage indices from D_1 to $D_{4.2}$ for each train considered.
$D_{1,str}$, D_2 and D_3 are the components of the damage on the straight sections while $D_{1,curv}$, $D_{4.1}$ and $D_{4.2}$ are the components of the damage on the curve sections.

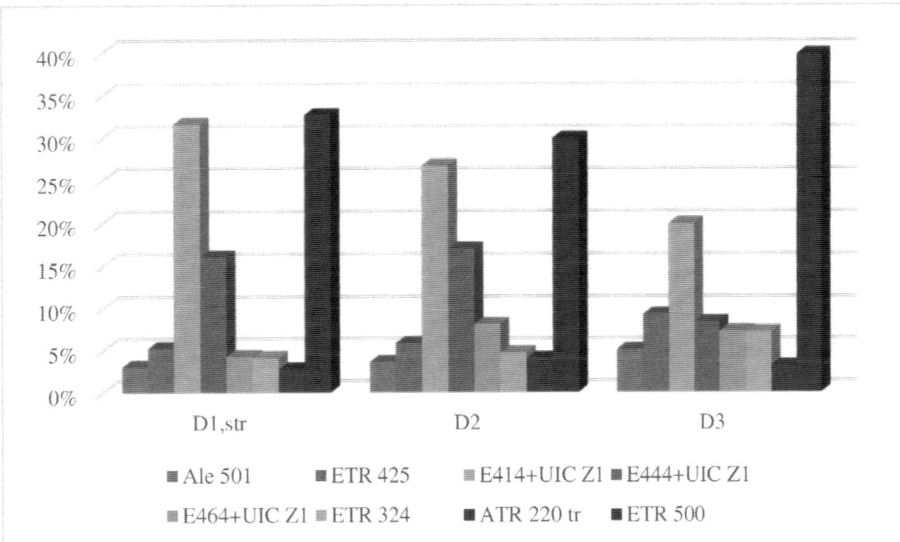

Figure 6. Damage on straight sections

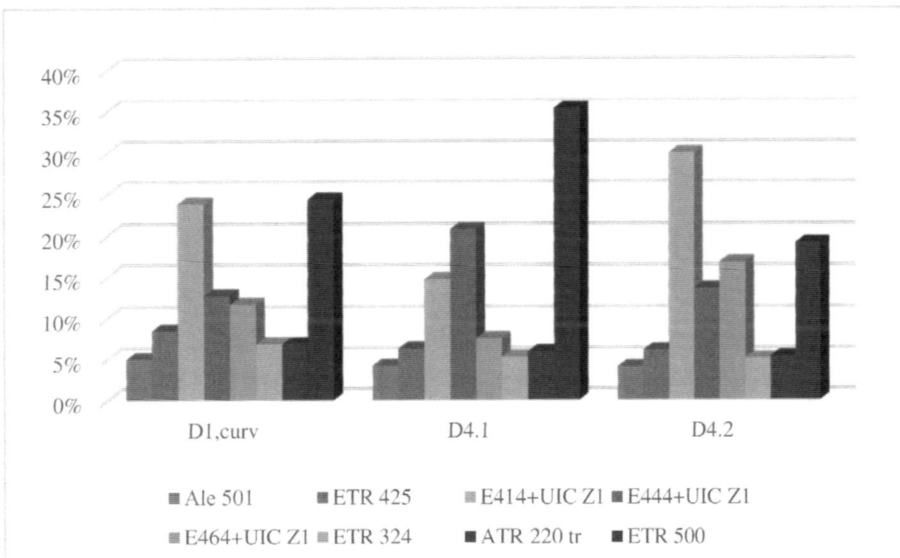

Figure 7. Damage on curve sections (r = 800 m)

The percentage values were calculated in relation to the total value of each damage index, caused by the transit of all trains on the entire line.
As can be seen, the most damaging vehicles are E414+UIC Z1 and ETR 500, which belong to high speed category. This can be due to the high mass and traction power that characterize them. For D_1,

and D_2,which depend by maximum dynamic load for single wheel, and for $D_{4.1}$, the E444+UIC Z1 is more damaging than other vehicles considered (except high speed vehicles). For D_3, all non-high speed vehicles cause almost the same damage. Finally, for $D_{4.2}$, the damage caused by E444+UIC Z1 and E464+UIC Z1 is comparable to the damage caused by high speed vehicles.

4. Conclusions

The study presented in this paper has been focused on the incidence of some types of rolling stocks on the track damage. The analysis of the specific contribution on damage by a vehicle type will be more important in order to modulate track access charges by vehicle type.

The simulation results will allow a dual purpose: on the class of train whose volumes are known on the line (in terms of trains*km or t*km) and to provide damage quantification, and therefore ordinary maintenance costs on the line, varying traffic volumes (train*km or t*km), train composition (% freight, regional, long haul) and vehicle parameters.

As written in paragraph 2, maintenance activities are associated with damage indices; if the maintenance costs are known, it is possible to associate them with the damage functions by the use of some calibration coefficients (k_i) [4]. Therefore, these coefficients allow to establish the costs directly related to the passage of vehicles on the railway infrastructure in order to calculate the direct costs related to the entire route considered.

References

[1] Smith, A., Iwnicki, S., Kaushal, A., Odolinski, K., Wheat, P. Estimating the relative cost of track damage mechanisms: combining economic and engineering approaches (2017) Proceedings of the Institution of Mechanical Engineers, Part F: Journal of Rail and Rapid Transit, 231 (5), pp. 620-636.

[2] Kuka, N., Verardi, R., Ariaudo, C., Pombo, J. Impact of maintenance conditions of vehicle components on the vehicle–track interaction loads (2018) Proceedings of the Institution of Mechanical Engineers, Part C: Journal of Mechanical Engineering Science, 232 (15), pp. 2626-2641.M. Bond, "A bear called Paddington", Collins, London, United Kingdom, 1958.

[3] NZV-BAV_TP2017_Anleitung EN Verschleissfaktor Fahrbahn Federal Office of Transport (BAV), 3003 Bern, Finance department Published on BAV website

[4] ERRI B176. Bogies with steered or steering wheelsets" European Rail Research Institute (ERRI), 1989

IOP Conf. Series: Materials Science and Engineering **922** (2020) 012007 doi:10.1088/1757-899X/922/1/012007

An Active Compliance Control for A Robotic Wound Closure System

Edward H. Currie[1], Y. M Zhao[1], Louis Kavoussi[2], Sina Y. Rabbany[1]

[1]DeMatteis School of Engineering and Applied Science, Hofstra University, Hempstead, New York
[2]Department of Urology, Long Island Jewish Medical Centre

Abstract. Most approaches to medical robotic compliance control rely upon forces, or torques, exerted on the mechanical arm links. The majority of robotic arms allow only position control, and orientation, of the end-effector. This paper uses a soft control technique to develop a position-correction, impedance method that leads to similar responses to the contact environment for simulation. The system is applied to control the contact force between a fixture and the skin in a robotic, vision-controlled, wound closure system. The experimental results show that the algorithm is successful in controlling contact force in fixture placement on different skin models.

1. Introduction

The importance of interaction control has been a subject of study in the last decades, resulting in advances in power electronics, computational power, and sensors [1]. The challenge with this approach is the contact's nonlinear behaviour, where classical techniques fall short [2]. Alternative approaches dealing with the transition from unconstrained-to-constrained motion have been studied including impedance control, has been proposed by Hogan [3]. This latter technique has received significant attention in recent years, as impedance control is has the potential to achieve the desired mechanical interaction within uncertain environments [4]. In the present context, impedance refers to the dynamic relation between the motion variables of a manipulator and the contact force. The primary benefit of impedance control is that it provides an effective way to simultaneously control the motion and contact force. This is paramount developing safe autonomous human-robot interactions. The implementation methods can be categorized into two main classes: a software-based approach [5-6] and a hardware-based approach [7-8]. Although many innovative applications use this technique, it is not a trivial task to achieve, therefore it is worthwhile to its corresponding algorithmic implementation.

Manual surgical procedures such as suturing can exhibit significant variations in technique and effectiveness due to the procedure's dependence on an individual manual dexterity. In order to optimize as well as standardize outcomes, interest in robotic wound closure systems has evolved. The concept of impedance control was introduced into medical robotics research and has gradually resulted in many specific implementation strategies [9]. Moreover, medical applications need to interact with the environment in such a compliant way that superior dexterity, over that of humans, is provided to the end-effector of a robotic arm. This paper presents a method to control the contact force between a fixture and the skin in a robotic wound closure system using a position-correction, impedance control, strategy organized as follows: Section 2 gives some essential explanations about the concept of impedance control. Position-correction methods to implement impedance control are introduced in

IOP Conf. Series: Materials Science and Engineering **922** (2020) 012007 doi:10.1088/1757-899X/922/1/012007

Section 3. The skin model, the interaction of fixture and wound skin, and a schematic diagram of position-correction impedance control are described in Section 4. The simulation and experimental results are presented in Section 5. Section 6 presents conclusions

2. Impedance control implementation

2.1. Impedance control dynamic equation.
Impedance control may be understood as a complex scheme behaving like a second-order, dynamic system. Figure 1 displays a simplified model of a mechanical assembly, with inertial, spring and damper components. Equation (1) describes the model's dynamics in the Laplace domain, where $X(s)$ is the displacement of the Tool Centre Point (TCP), $F_{int}(s)$ is the interaction force with the environment, M is the inertial component, B is the energy dissipation term from the viscous damper, and K is the stiffness of the equivalent system.

$$X(s) = \frac{F_{int}(s)}{Ms^2+Bs+K} = G(s)F_{int}(s) \tag{1}$$

2.2. Solution by discrete-time convolution.
Discrete-time convolution is an operation on two discrete-time signals defined by the integral

$$(f * g)(n) = \sum_{k=-\infty}^{\infty} f(k)g(n-k) \tag{2}$$

for all signals f, g defined on Z. The above operation has proven to be particularly useful in the study of linear, time-invariant systems. In order to see this, consider a linear time-invariant system H with a unit impulse response h. Given a system input signal x, we would like to compute the system's output signal $H(x)$. First, we note that the input can be expressed as the convolution:

$$x(n) = \sum_{k=-\infty}^{\infty} x(k)\delta(n-k) \tag{3}$$

Where $\delta(n-k)$ is unit sample sequence or unit impulse or Kronecker delta function, i.e.,

$$\delta(n-k) = \begin{cases} 1 & n = k \\ 0 & n \neq k \end{cases} \tag{4}$$

Since $H\delta(n-k)$ is the shifted unit impulse response $h(n-k)$, this gives the result

$$Hx(n) = \sum_{k=-\infty}^{\infty} x(k)h(n-k) = (x * h)(n) \tag{5}$$

Hence, convolution has been defined such that the output of a linear, time-invariant system is given by the convolution of the system input with the system's unit impulse response.

3. Skin modelling and contact interaction

3.1. Skin modelling. Most materials can be considered linear if the strain and stress remain small. If the strain remains linear, the mechanical behaviour is considered linear. Such linear behaviours can then modelled by parallel and/or a serial combination of linear springs (i.e., springs with a constant stiffness coefficient) and linear dashpots (with a constant damping coefficient), e.g., linear viscoelastic bodies such as a generalized Kelvin-Voigt body or a generalized Maxwell body as presented in Figure 1 (a). Figure 1 (b) shows the interaction between the skin and the end-effector.

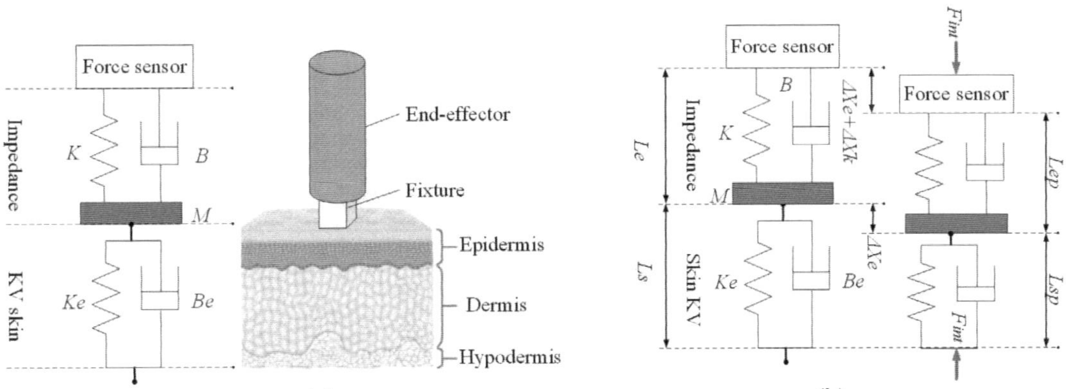

(a) (b)

Figure 1. Skin modelling and contact impedance control dynamic model: (a) *M-B-K* impedance parameters; *Ke-Be* skin model parameters, (b) Representing interaction analysis of skin and robot end-effector with contact force $F_{int}(t)$.

3.2. Schematic diagram. The control schematic diagram of position-correction impedance control is illustrated in figure 2.

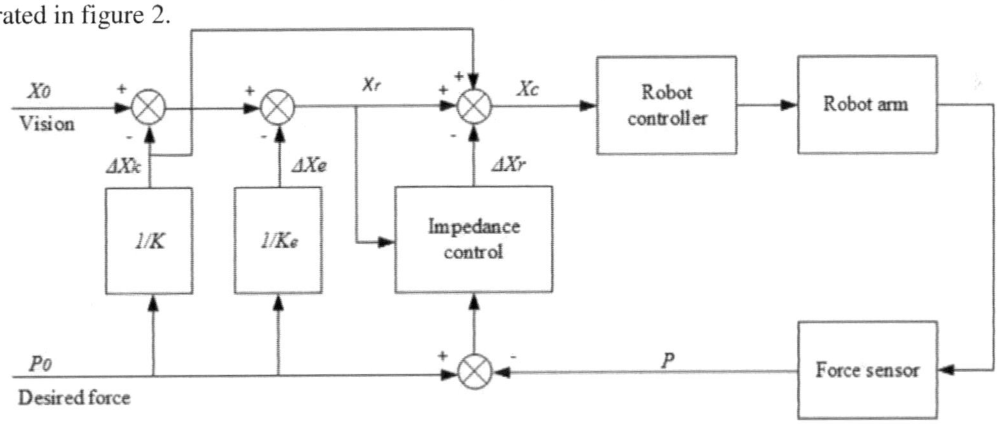

Figure 2. Diagram of position-based impedance control for the force between skin and robot arm

Figure 3. System configuration:
1 - Stereo camera, 2 - Robotic arm,
3 - Force sensor,
4 - Laser projector,
5 - Wound surface.

The correction to the robot's position, X_c, is the input to the robot controller in Figure 2. P_0 is the desired contact force and X_0 is the desired position of the fixture on the skin's surface, which is determined by a vision-guiding system.

4. Simulation and experiment results

4.1. Experimental setup and e parameters. The simulations and experiments validated the robotic arm position-correction impedance control strategy. The simulations were carried out in a Simulink environment (MATLAB 2018a). The experiments were performed on the robotic, vision-control wound closure system shown in Figure 3. The robotic, vision-control wound system included a Mitsubishi RV-2SDB robotic arm, fixture applier, force sensor installed inside fixture applier, wound model, stereo camera, and a laser pattern projector. For each task, the impedance parameters used were M = 10 kg, B = 18 $N.s/m$, and K = 25 N/m.

4.2. Validation of contact force

Three skin models were studied for tracking the desired force trajectory using the proposed approach. The skin model parameters are list in Table 1.

Table 1. Skin model parameters.

Skin model	Parameters	
	$Be\ (N.s/m)$	$Ke\ (N/m)$
A	0.01	10
B	26	625
C	11	576

Determination of the unit impulse response for a discrete-time $h(k)$ is illustrated in figure 4. To prevent the overshoot of the response in fixture placement on the skin's surface, the impedance of the robotic arm functions as a second-order linear system with critical damping which provides the quickest approach to zero amplitude for a damped oscillator. The robot's end-effector critical damping coefficient is given by: $B = 2 \cdot \sqrt{MK}$

If $M = 100\ (kg)$, $K = 100\ (N/m)$, then $B = 200\ (N.s/m)$. A unit impulse was used for the position-correction impedance control of the robot end-effector and is shown in figure 4 (a). The response to the unit impulse for a discrete-time $h(k)$ of the robotic arm in critical damping is shown in figure 4 (b).

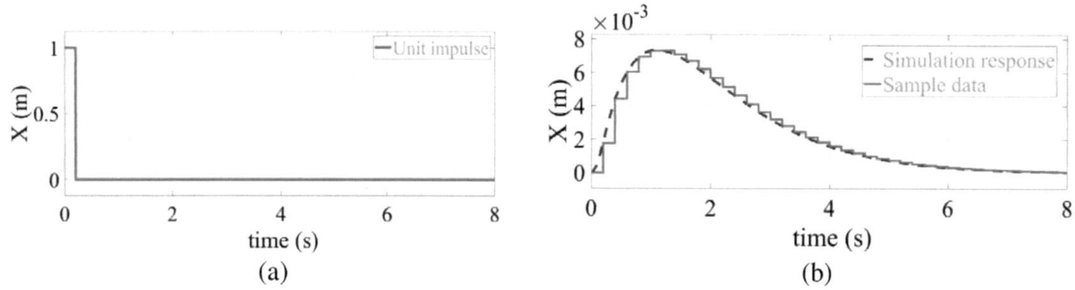

Figure 4. A unit impulse was used for position-correction impedance control and the response to a unit impulse for a discrete-time $h(k)$ of the robotic arm in critical damping.

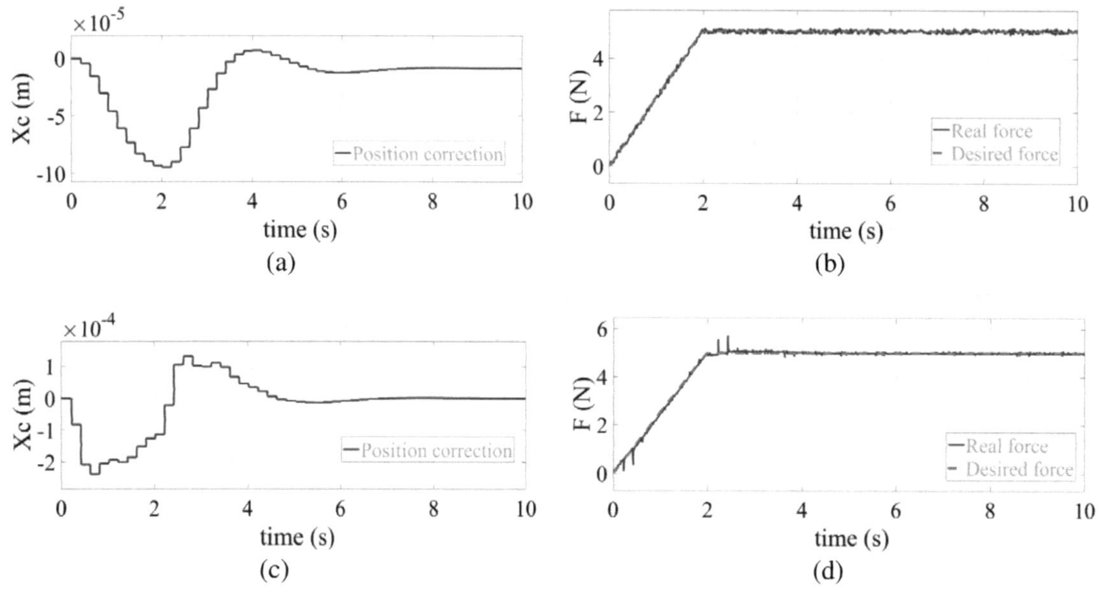

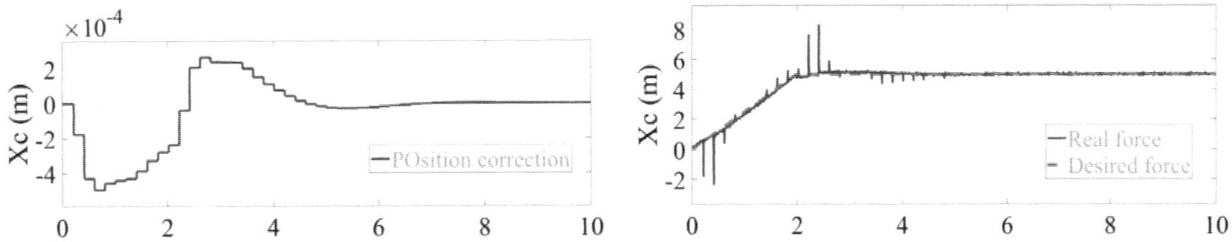

Figure 5. Experimental results for contact force tracking by position-correction impedance control: (a) Correction of robot end-effector of case A, (b) Contact force tracking of case A, (c) Correction of robot end-effector of case B, (d) Contact force tracking of case B, (e) Correction of robot end-effector of case C, (f) Contact force tracking of case C.

The experimental results are shown in figure 5. The corrections of the robot end-effector position are plotted in figure 5 (a), (c), and (e), respectively. Contact force tracking of the desired force trajectory is given in figure 8 (b), (d), and (f), respectively. The results demonstrate that the proposed position-correction impedance control can accurately track the contact force between the fixture and the skin surface. It can also track the reference contact force when the parameters of the skin model change. This implies that position-based impedance can be used for different skin models without changing the control parameters. The tracking errors for the reference force trajectory were measured and the errors evaluated in terms of the average errors (*AEs*) and standard deviation (SD), in the X-direction. The errors were computed, and the results are listed in Table 2.

Table 2. Formatting sections, subsections and subsubsections.

Case	Tracking error	
	AEs (N)	*SD (N)*
A	0.011	0.156
B	0.046	0.026
C	0.034	0.032

Table 2 shows that the softer the skin becomes, the smaller the *AEs* in X. It is also demonstrated that the proposed method can track the contact force between the fixture and the skin surface with 0.34 (N) and AEs within 3 (sec).

5. Conclusions

A contact force, tracking method based on the position-correction approach is proposed in this paper. The simulation and experimental results of contact force tracking on the skin models with different parameters or on different part of body was studied utilizing a robotic, vision-controlled, wound closure system. The results show that the proposed method, based on a position-correction of the robot end-effector, can accurately track the desired contact force trajectories between the end-effector and the skin models that have different parameters.

6. References

[1] L. Villani and J. de Schutter, *Force Control, in Springer Handbook of Robotics*, B. Siciliano and O. Khatib, Eds. Berlin, Heidelberg: Springer Berlin Heidelberg, 2008, pp. 161184.

[2] N. Hogan, "Adaptive Control of Mechanical Impedance by Coactivation of Antagonist Muscles," *IEEE Transactions on Automatic Control*, vol. 29, pp. 681-690, Aug. 1984.

[3] N. Hogan, "Impedance Control: An approach to manipulation: Parts IIII," *J. Dyn. Syst. Meas. Control. Trans. ASME*, vol. 107, no. 1, pp. 17, 1985.

[4] P. Song, Y. Q. Yu and X. P. Zhang, "Impedance Control of Robots: An Overview," *IEEE International Conference on Cybernetics, Robotics and Control*, pp. 51-55, July 21-23, 2017.
[5] R. J. Anderson and M. W. Spong, "Hybrid impedance control of robotic manipulators," *IEEE Journal on Robotics and Automation*, vol. 4, pp. 549-556, Oct. 1988.
[6] B. Heinrichs, N. Sepehri and A. B. Thornton-Trump, "Position-based Impedance Control of an Industrial Hydraulic Manipulator," *IEEE Control Systems*, vol.17, pp.46-52, Feb. 1997.
[7] J. W. Sensinger and R. F. f. Weir, "Unconstrained Impedance Control Using a Compact Series Elastic Actuator," *Proc. of 2nd IEEE/ASME International Conference on Mechatronics and Embedded Systems and Applications*, pp. 1-6, Aug. 2006.
[8] B. Vanderborght et al., "Variable Impedance Actuators: Moving the Robots of Tomorrow," *Proc. of IEEE/RSJ International Conference on Intelligent Robots and Systems*, pp. 5454-5455, Oct. 2012.
[9] Y. H. Tsoi and S. Q. Xie, "Impedance Control of Ankle Rehabilitation Robot," *Proc. of IEEE International Conference on Robotics and Biomimetics*, pp, 840-845, Feb. 21-26, 2009.
[10] Y. C. Fung, *Biomechanics: Mechanical Properties of Living Tissues*, 2nd ed: Springer-Verlag New York, Inc., 1993.

Acknowledgements

This research was supported by the research fund from Northwell Health Hospital and Hofstra University.

IOP Conf. Series: Materials Science and Engineering **922** (2020) 012008 doi:10.1088/1757-899X/922/1/012008

Design and development of an automated fluid management system for endoscopy aided gynaecological surgeries

S P Hettiarachchi[1,2], G C K Pathirana[1], M R Perera[1], Y W R Amarasinghe[1,2], M A M M Jayawardane[2,3] and V T Dau[4]

[1]Department of Mechanical Engineering, University of Moratuwa, Katubedda 10400, Sri Lanka
[2]Centre for Advanced Mechatronic Systems – Accelerating Higher Education Expansion and Development (AHEAD) project, University of Moratuwa
[3]Faculty of Medical Sciences, University of Sri Jayewardenepura, Nugegoda 10250, Sri Lanka
[4]School of Engineering and Built Environment, Griffith University, Queensland, Australia

Abstract. In this research, a novel Automated Fluid Management System (AFMS) was designed, analysed and fabricated for endoscopy aided gynaecological surgeries in order to prevent complications arise due to fluid overload. The proposed system consists of a peristaltic pump, suction mechanism and fluid measurement system integrated with a graphical user interface which can also control and monitor through a wireless network system. The device serves functionalities such as regulating saline flow rate, limiting saline suction rate, setting saline absorption limits for different patients and indicating real-time fluid deficit. This paper discusses designing of a non-contact pump for gynaecological surgeries, development of the control system for AFMS and the comparison of experimental results of the fabricated non-contact pump with theoretical pump characteristics.

1. Introduction

Endoscopy is one of the minimal invasive technique used to observe internal organs or tissues in detail. Those endoscopic approaches are used to visualise the uterine cavity for diagnosis and treatment of intrauterine pathology. During this procedure, either fluid or carbon dioxide can be used as distention medium in order to achieve better observations [1]. As bleeding occurs in operative procedures, it results in obstructing the visualisation of the uterine cavity. Therefore, most surgeries are carried out with fluid media such as saline because it provides a clear picture by continuous irrigation. During these gynaecological surgeries, absorption of a large volume of distension solutions can lead to serious complications arise due to fluid overload which will cause for conditions such as electrolyte imbalance, pulmonary oedema and cardiac failure [2]. Since saline is considered as one of the isotonic solutions, it is recommended that absorption of saline should be within the range of 1500-2500 ml for elderly or those women with co-morbid conditions and healthy women of reproductive age [3]. Therefore, a fluid management system with real-time fluid deficit indication is a necessity for the surgeons these days.

As the distension medium is pumped into the human body, it should be ensured that it is not contaminated in any way. In order to avoid any contact between pump parts and the fluid, a non-contact type pump should be used in the fluid management system. Peristaltic pumps are one of the main non-contact type pumps which prevent any contamination of the transported fluid [4]. These

pumps contain two or more rollers which press the flexible tube to the manifold where the pump is placed [5]. Therefore, developing a peristaltic pump with the capability to alter the flowrate as required for different surgeries is another necessity in a fluid management system.
This paper proposes a novel automated fluid management system for endoscopy aided gynaecological surgeries with design, analysis and development of a peristaltic pump, a suction mechanism with a diaphragm pump which can limit the suction flowrate, fluid volume measurement method with infrared sensor and graphical user interface to monitor and control the system. Further, the device has the capability to connect with pre-defined Wi-Fi network and provide remote access of the system as well. As a solution to minimise the error arise due to blood in the measurement of sucked out fluid a method is proposed to evaluate the blood volume using a colour code for different blood concentrations mixed in a saline solution.

2. Proposed design and working principle of the fluid management system

In the AFMS saline is fed through the non-contact pump with the help of gravitational effect. The flowrate control of the non-contact pump is regulated by the user through a graphical user interface or using wireless communication. Thereby, saline flows into the body cavity of the patient. After that, the saline inserted into the body is removed using a diaphragm pump into the fluid collector, where the output fluid is measured using sharp infrared sensor while the deficit is displayed real-time on the graphical user interface. Figure 1 shows the working principle of the proposed fluid management system graphically.

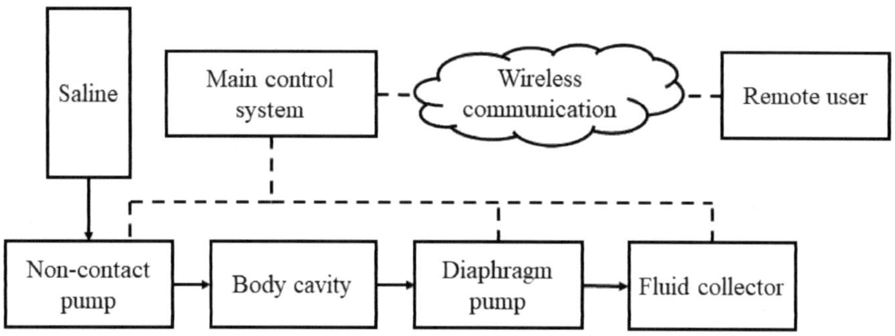

Figure 1. Graphical representation of the working principle of the proposed fluid management system

In the aspect of developing the fluid management device, the prominent focus was given to design and development of a non-contact pump for pumping the saline into the body cavity without any contamination and to the development of a control system to integrate all the components of the device.

2.1. Design and development of a non-contact pump

A non-contact pump is essential for the device as it is recommended that the saline should not be contaminated with any of the pump parts or abrasive media. After a careful study of existing non-contact type pumps, it was identified that peristaltic pumps are widely used in medical applications due to its reliability of providing contamination-free liquids with better accuracy of flow control and suitable for these kinds of applications [6].
A peristaltic pump is a type of positive displacement pump where the fluid is contained within a flexible tube fitted in a circular pump casing. A rotor with several rollers, wipes or shoes attached to the external circumference compresses the flexible tube [7]. Major design considerations of the peristaltic pump were continuity and flowrate. It was identified that the maximum flowrate requirement of the surgeons for this particular application as 1000 ml/min. The pulsating effect of a peristaltic pump is unavoidable due to its nature of the operation. Therefore, in order to obtain a nearly continuous flow, the pump the occluded length due to the rollers should be minimised. To minimise this length the only option is to increase the number of rollers in a fixed rotor. Because of that total of

six rollers were used in the design of the peristaltic pump. The detailed representation of the designed non-contact pump with its components is shown in figure 2.

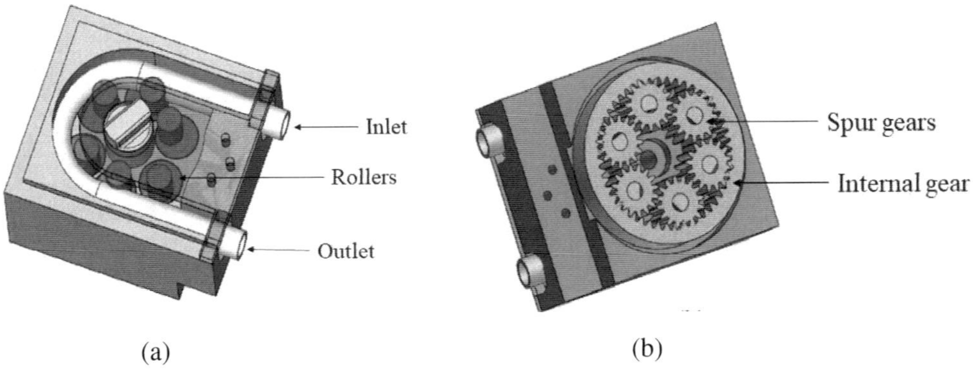

Figure 2. 3D model of the pump mechanism (a) Top view (b) Bottom view

The theoretical flow rate of a peristaltic pump is given in equation (1).

$$\text{Theoretical flow rate} = A \times L \times N \times \omega \text{ ml/min} \quad (1)$$

Here 'A' is the area of the tube, 'L' is length occluded by rollers, 'N' number of rollers and 'ω' is the angular velocity. The tube diameter of the design used is 5 mm and the radius of the curvature of the rotor was 22.3 mm. Due to the gravitational effect on the saline pumping used for gynaecological surgeries, it was observed that an average of 82 ml/min can be achieved at that specific height. Therefore, equation (1) can be modified to equation (2) in order to provide an understanding of how the flowrate of the design peristaltic pump varies with the angular speed of the motor.

$$\text{Flow rate} = 1.834 \times \omega + 82 \text{ ml/min} \quad (2)$$

Equation (2) can be used as the pump characteristic equation for the designed peristaltic pump. According to that, to obtain a maximum flow rate of 1000 ml/min the pump should have an angular velocity around 500 rpm. A gear mechanism with a gear ratio of 3.43 was implemented to the design to obtain the required angular velocity with high torque motor.

All the components of the proposed design were fabricated using 3D printing technology with acrylonitrile butadiene styrene material.

2.2. Development of the control system for AFMS

The control system of the device includes actuator and sensor assembly, main control system, graphical user interface and wireless sensor network system. Integration of the system components with the control system is shown in figure 3.

IOP Conf. Series: Materials Science and Engineering **922** (2020) 012008 doi:10.1088/1757-899X/922/1/012008

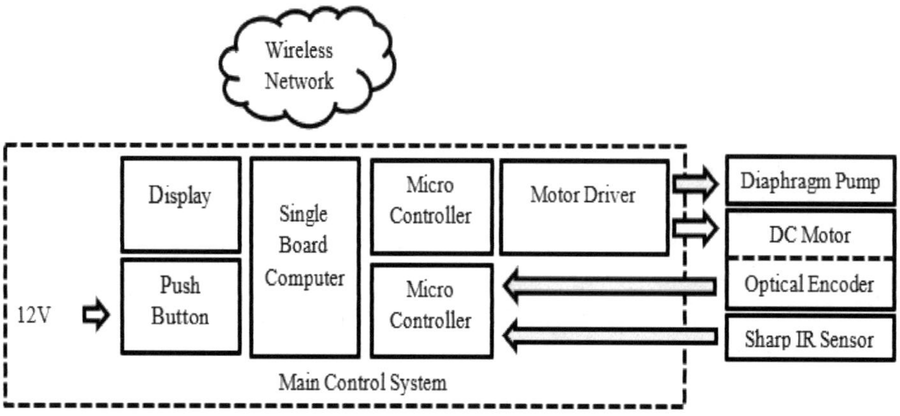

Figure 3. Graphical representation of the proposed control system

2.2.1. Main control system.

The main control system of the device consists of HDMI touch display, power on/off button, a single board computer, two microcontrollers and a motor drive. Single board computer helps to process all the data that are collected from sensors and inputs and controls entered from the user through the display. The software for the AFMS was developed using Python programming language on Raspberry Pi 3 single-board computer. Serial communication is used to distribute the collected data to relevant microcontrollers. Two Arduino microcontrollers were used for reading the sensor values and to reduce the processing time of the graphical user interface operation. The motor drive has the ability to change the speed of the DC motor and it has the controlling of the suction flow rate of the diaphragm pump which has a set maximum from the user.

2.2.2. Sensors and actuators.

A DC gear motor with inbuilt encoder, a DC diaphragm pump and a sharp infrared sensor were used as actuators and sensors. The DC gear motor is connected to the fabricated peristaltic pump which has an inbuilt encoder to obtain motor feedback to the controller. Using Proportional-Integral-Derivative (PID), it helps to maintain a constant rpm for the motor resulting in a constant flow rate for saline pumping. Diaphragm pump was used as the suction pump to remove saline from the body cavity. This removed fluid is collected to a container which has a constant cross-section area. To obtain the volume of sucked out fluid sharp infrared sensor was integrated after calibrating it using the floating duck mechanism.

2.2.3. Graphical user interface.

Graphical user interface development is required in order to provide external inputs to the control the device by authorised personnel. User has the ability to choose the required surgical procedure, where predefined values for the flow rates and limits for that certain procedure can alter manually depending on the patient. After proceeding and initiating the suction process the saline deficit value updates in real-time. The device provides a warning when the saline deficit reaches to the maximum allowable absorption volume. The developed graphical user interface is shown in figure 4.

2.2.4. Wireless communication.

The device has the ability to connect with a predefined Wi-Fi network allowing remote access of the system to the user. User can alter or examine the set parameters of the interface providing flexibility in controlling and monitoring of the device remotely.

IOP Conf. Series: Materials Science and Engineering **922** (2020) 012008 doi:10.1088/1757-899X/922/1/012008

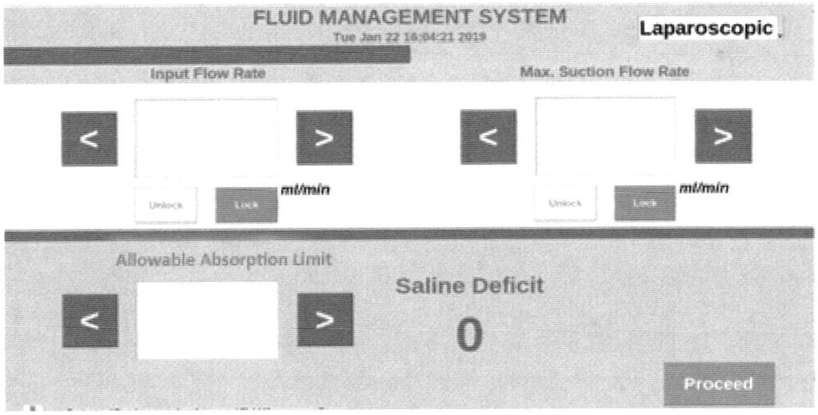

Figure 4. The graphical user interface of the device

3. Results and discussion

The testing of the fabricated pump was conducted in order to obtain the theoretical and experimental pump characteristic curves. In the experiment, DC gear motor voltage values were mapped to the voltage values (0-12V to 0-255) and with its angular velocity allowing it to change the flow rate of the pump accordingly. Following results in table 1 were obtained for varied rpm values and pump characteristic curve obtained from equation (2) with the experimented results obtained were plotted as shown in figure 5.

Table 1. Theoretical and experimental flow rates for fabricated peristaltic pump.

Voltage mapped value	Motor rpm	RPM of rollers	Flow rate	
			Theoretical (ml/min)	Experimental (ml/min)
255	160	548.80	1088.50	1053
220	138.04	473.47	950.35	918
200	125.49	430.43	871.41	845
180	112.94	387.39	792.47	780
160	100.39	344.35	713.53	682

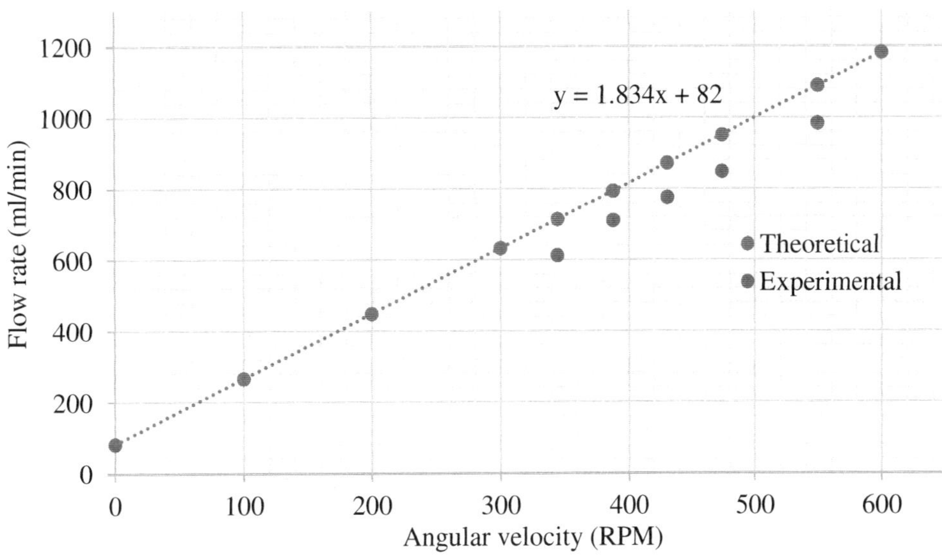

Figure 5. Pump characteristic curve and experimented results

As per the experiment results, it has an error of 11.23% with the theoretical values. This variation is mainly due to the precision of manufacturing of the pump. Since all the components of the pump were 3D printed, obtaining a better surface finish to precise control of the mechanism is harder.
With compared to the volume of saline inserted into the body cavity, it was identified that the amount of blood coming out with the saline is negligible. But it is inevitable that the sucked-out fluid containing a certain amount of a blood volume and therefore, a colour code was introduced in order to minimize the error in calculating absorption of saline volume. In preparing the colour code a known amount of blood was mixed with a known amount of saline and observed the colour variation with respect to the concentration of the blood volume. Therefore, the user can have an idea about the exact saline volume that has been absorbed into the body thus can prevent fluid over absorption complications.

4. Conclusion

An automated fluid management system with a novel design of a peristaltic pump, a fluid measuring system and control mechanism was developed under this research for endoscopy aided gynaecological surgeries. Results obtained from the experiment showed a deviation from the theoretical pump characteristic due to the quality of the fabricated pump. The deviation can be minimised by developing the pump using advanced manufacturing technologies. All the functionalities of the device can be accessible via developed wireless network system providing convenient for surgeons in monitoring and controlling of the device. In order to minimise the error, arise due to blood containing in the sucked-out fluid mixture, a colour code was introduced for various blood concentrations. In future, the device can be further optimised with image processing instead of colour code for accurate measurement of the blood volume in the sucked-out fluid mixture.

5. References

[1] Brusco G F, Arena S and Angelini A 2003 Use of carbon dioxide versus normal saline for diagnostic hysteroscopy Fertility and Sterility 79 993-97
[2] Varol N, Maher P, Vancaillie T, Cooper M, Carter J, Kwik A, Pesce A and Reid G 2002 A literature review and update on the prevention and management of fluid overload in endometrial resection and hysteroscopic surgery Gynaecological Endoscopy 11 19-26
[3] Jensen F W, Vredevoogd C B, Van Ulzen K, Hermans J, Trimbos J B and Trimbos-Kemper T C M 2000 Complications of Hysteroscopy Obstetrics & Gynecology 96 266-70
[4] Klespitz J and Kovács L 2014 Peristaltic pumps – a review on working and control possibilities IEEE 12th Int. Symp. on Applied Machine Intelligence and Informatics Herl'any pp. 191-4
[5] Klespitz J and Kovács L 2013 Identification and control of peristaltic pumps in hemodialysis machines IEEE Int. Symp. On Computational Intelligence and Informatics Budapest pp. 83-86
[6] Todorov V, Aprahamian B and Tahrilov H 2016 Applicability of peristaltic pumps in installations for vegetable oils processing 19th Int. Symp.on Electrical Apparatus and Technologies Bourgas pp. 1-4
[7] Dhumal S R and Kadam S S 2012 Design and development of rotary peristaltic pump Int. Journal of Science and Advanced Technology 2 4

Acknowledgement
The authors would like to express their gratitude to the Department of Mechanical Engineering, University of Moratuwa and Accelerating Higher Education Expansion and Development (AHEAD) project of Centre for Advanced Mechatronic Systems, University of Moratuwa for their valuable advice and guidance towards the success of the research.

IOP Conf. Series: Materials Science and Engineering **922** (2020) 012009 doi:10.1088/1757-899X/922/1/012009

Vehicle Positioning System using V2X that Combines V2V and V2I Communications

Geon-Hee Kim[1],Dong-Sung Pae[1],Woo-Jin Ahn[1],Koung-Suk Ko[1],Myo-Taeg Lim[1]*, Tae-Koo Kang[2]*

[1] School of Electrical Engineering, Korea University, Seoul, Republic of Korea
[2] Department of Information and Telecommunication Engineering, Sangmyung University, Cheonan, Republic of Korea

Corresponding author : mlim@korea.ac.kr, tkkang@smu.ac.kr

Abstract. In this paper, we present a V2X (Vehicle to Everything) vehicle positioning system environment using V2V (Vehicle to Vehicle) and V2I (Vehicle to Infrastructure) communication server. With the growth of the 5G network technology, using V2X which exchanges information, connected car recently became an important area of research in autonomous driving. However, it takes too much risk to test autonomous driving in a real-world experiment and it also have difficulties such as road infrastructure or 5G communication system. Regrading autonomous driving research, we established V2X environment in a small RC car, similar to a vehicle environment. Our proposed Vehicle Positioning System environment uses a ceiling camera, V2V, and main server, V2I, which plays the same role as GPS and 5G in the real-world. Combining the V2I and V2V from the camera of each other car using Kalman Filter we present a vehicle positioning system.

1. Introduction

Recently, autonomous driving technology is rapidly being studied [1][2][4][6]. Its most important criteria in this technology is safety and support. Typical autonomous driving techniques such as sensor-based ADASs (Advanced Driver Assistance Systems) and connection-based V2X (Vehicle to Everything) are emerging as key technologies in autonomous driving system. The ADAS is equipped with state-of-the-art sensors such as cameras, radar, LiDAR (Light Detection and Ranging) and ultrasonic which indicates the risk of accidents during driving or parking [3][4]. However, ADAS sensors provide a short detection range and limited detection of situations behind obstacles and their constraints which can only be used within non blind spot area. This limitation can be covered with V2X system, which tells that key technology for autonomous driving, in particular, is the V2X system [5][6].

The V2X also supports autonomous driving in combination with "precision measurement technology," that measures the location of cars more precisely and informing the drivers and cars. This not only allows autonomous driving cars to reduce unexpected traffic accidents in unexpected situations that people or sensors cannot recognize, but also reduces car gas emissions by easing the traffic jam. As shown in Figure 1, V2X consists of V2V (Vehicle to Vehicle), V2I (Vehicle to infrastructure), and V2P (Vehicle to Pedestrian) system, etc. V2X system means communication with the vehicle and all 'X' and the V2X system is a technology that can improve driving stability and

IOP Conf. Series: Materials Science and Engineering **922** (2020) 012009 doi:10.1088/1757-899X/922/1/012009

convenience using the different infrastructure [7][8]. Therefore, V2X system prevents traffic accidents by knowing vehicle and pedestrian information.

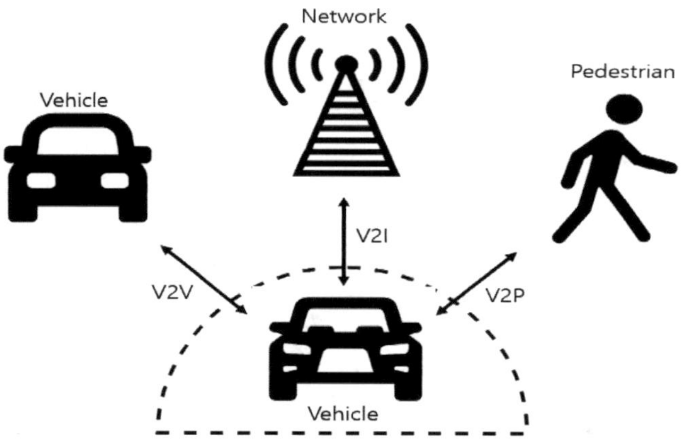

Figure 1. Types of V2X communication

V2V refers to a technology that transmits and receives information wirelessly between vehicles. Information can be exchanged between vehicles to prevent sudden accidents by exchanging location and speed information [9]. V2I refers to the technology that sends and receives information wirelessly between vehicles and road infrastructure. This refers to a technology that collects driving information from vehicles by installing base stations around roads and provides traffic information to vehicles by analysing it from a central server [10]. Through this process, traffic congestion or traffic accidents can be prevented in advance. Therefore, The V2X system, which communicates with road infrastructure and other vehicles while driving, will allow for more efficient autonomous driving.

However, autonomous driving tests using real vehicles can lead to dangerous accidents if an accident occurs due to an algorithm error or sensor error during autonomous driving. Therefore, we constructed a V2X environment using V2V and V2I system for small RC car-type experimental environment. In addition, Vehicle Positioning System was tested on the vehicle after setting up the V2X System environment using Kalman Filter. Section2 indicates the configuration of the Vehicle Positioning System by the V2X system. Since the system was then configured, Section 3 is shown for the experiment. Section 4 consists of a conclusion.

2. Vehicle Positioning V2X System

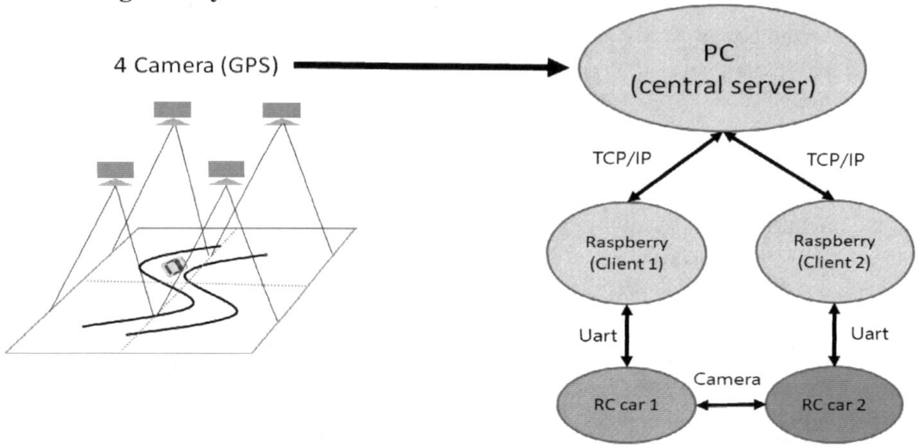

Figure 2. V2X System configuration

This chapter explain how V2V systems and V2I systems is configured, uses the data obtained from V2V and V2I systems, and explain the Vehicle Positioning V2X System using Kalman Filter from the system model on Newton's Law. We used Kalman Filter to use vehicle information from V2X for the system model. In [11][12][13], Kalman Filter are used to predict the trajectory of objects moving in 3D, 2D coordinates. Kalman Filter requires a kinematic model that may be inaccurate due to assumptions such as constant velocity or acceleration models and does not take into account factors such as wind resistance that will affect the model. The Kalman Filter also uses a measurement model that requires measurement data for the position, speed and acceleration of the vehicle to be measured using the camera.

Figure 2. shows the overall configuration of our proposed V2X system. The V2I system locates RC cars through four cameras and the central server receives the location information of the vehicle from the V2I system. The V2V system interacts with the central server, sharing the information from the vehicle's camera. Raspberry Pi is used in RC car to communicates information with the Central server through TCP/IP communication. The vehicle's location information from the V2I system and V2V system on the central server is noisy, making it difficult to pinpoint the exact location. Therefore, Kalman Filter is used in the V2X system to measure the location of the vehicle using the information received from the two systems.

2.1. V2V System

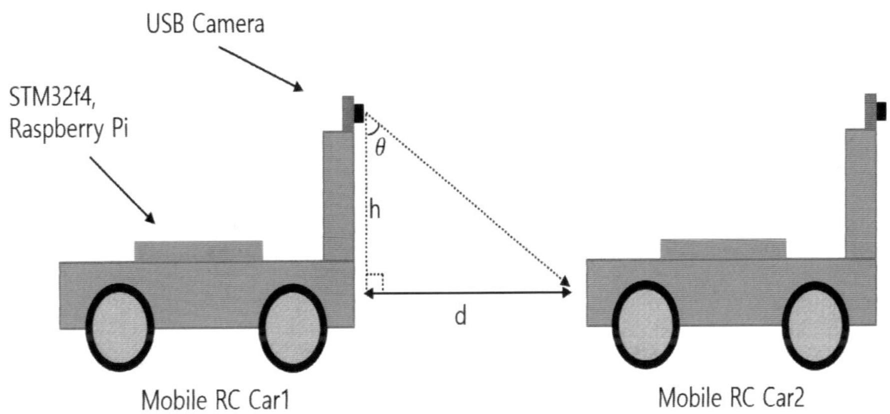

Figure 3. V2V System configuration

$$d = h \tan \theta \tag{1}$$

In this paper, V2V system includes camera sensors which provide multiple RGB data and helps predicting distances from obstacles. Camera is placed in front of the RC car, as shown in Figure 3. The distance can be predicted as shown in equation (1) for the front vehicle recognized by the camera. In other words, the V2V system communicates information between vehicles, so location and distance information can be known. However, the V2V system should incorporate the V2I system because it can predict the distance to a single-vehicle but does not know the exact information about a number of obstacle vehicles.

2.2. V2I System

The V2I system has difficulty to experiment with real vehicles. So, we did an experiment in RC car environment. Four cameras were installed on the ceiling to create an environment where RC car location information can be received like actual GPS information. Therefore, the central server can know the location of the vehicle. In addition, the central server can check the speed and direction of the vehicle through TCP/IP communication with the RC Car. Central server can provide autonomous driving path information to other vehicles with information from integrated vehicles.

2.3. Vehicle positioning using Kalman Filter

Central server has information received from the V2V system and provides location information for other vehicles. Information on vehicles appearing suddenly or disappearing can be transmitted to vehicles via the central server. After integrating V2I system and V2V system information, vehicle positioning will be able to accurately locate using Kalman Filter using a system model based on Newton's Law. Before using Kalman Filter to track an object, it needs to model the movement of this object. Although, the movement cannot be modelled accurately, it can have an approximation model of it. Assuming that the motion on the x-axis and y-axis and angular rotation around z-axis are uncorrelated. Measuring the position of the RC car every ΔT seconds, the system equation can be expressed as below formula based on Newton's Law. We can express the system in state equation as the follows [14]:

$$X(k+1) = FX(k) + Gu(k) + w \tag{2}$$

$$Z(k+1) = HX(k+1) + v \tag{3}$$

where,

$$X = \begin{bmatrix} x(k) & y(k) & \theta(k) & V_x(k) & V_y(k) & \omega(k) \end{bmatrix}^T \text{ and } u = \begin{bmatrix} a_x(k) & a_y(k) & \alpha(k) \end{bmatrix}^T,$$

$$F = \begin{bmatrix} 1 & 0 & 0 & \Delta T & 0 & 0 \\ 0 & 1 & 0 & 0 & \Delta T & 0 \\ 0 & 0 & 1 & 0 & 0 & \Delta T \\ 0 & 0 & 0 & 1 & 0 & 0 \\ 0 & 0 & 0 & 0 & 1 & 0 \\ 0 & 0 & 0 & 0 & 0 & 1 \end{bmatrix}, \quad G = \begin{bmatrix} \frac{\Delta T^2}{2} & 0 & 0 \\ 0 & \frac{\Delta T^2}{2} & 0 \\ 0 & 0 & \frac{\Delta T^2}{2} \\ \Delta T & 0 & 0 \\ 0 & \Delta T & 0 \\ 0 & 0 & \Delta T \end{bmatrix}, \quad H = \begin{bmatrix} 1 & 0 & 0 & 0 & 0 & 0 \\ 0 & 1 & 0 & 0 & 0 & 0 \\ 0 & 0 & 1 & 0 & 0 & 0 \end{bmatrix},$$

where,

x, y are position of the car, θ represent the angle, v_x, v_y are velocity of the car, and ω is angular velocity. a_x, a_y are acceleration, α is angular acceleration. and w, v are measurement noise.

3. Experiment

As shown in Figure 4., the experimental platform for RC car 1 and 2. Experimental RC cars were built using STM32F407 board, Raspberry Pi and RGB cameras (Microsoft HD-3000). The V2V System observes RC car 2 through the camera of RC car1 and measures the distance from RC car 2 as shown in Figure 5.

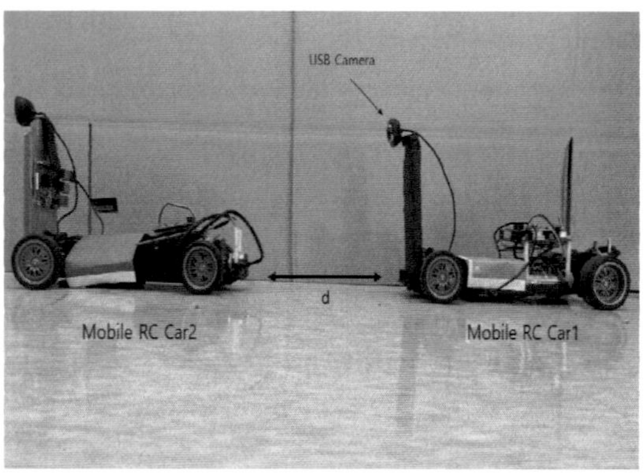

Figure 4. RC Car 1, 2 experimental platforms

Figure 5. V2V Communication

Figure 6 shows a V2I system that measures location information for RC car1 and 2. The V2V system connected four cameras to create a 640cm*480cm map and use four cameras to calibrate the camera and measure the location of RC car 1 and 2. Figure 7 shows the distance over time to RC car 1 and 2. also, Figure 7 shows Meaurement1 positioning results for the V2V The results show that there is a lot of noise. Similarly, Meaurement2 shows the positioning results that there is a lot of noise in the V2I system. Using the Kalman Filter based on two result values, the filtering values can be viewed as red lines and comes out more perfectly without any noise.

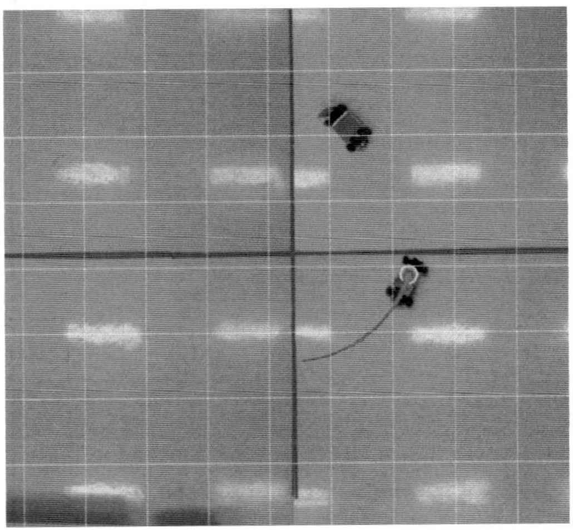

Figure 6. V2I Communication

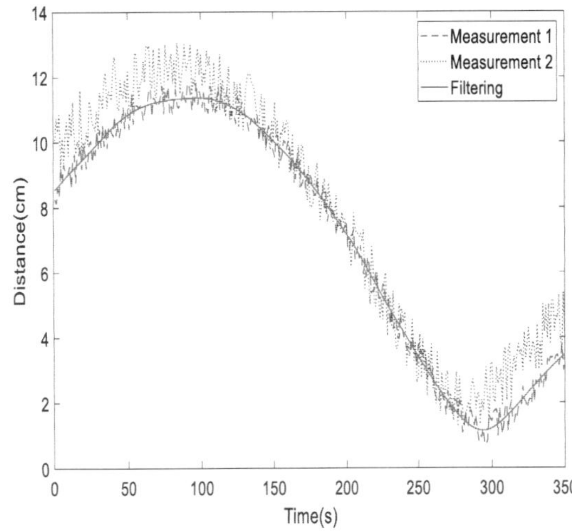

Figure 7. Vehicle Positioning results

4. Conclusions

In this paper, we proposed autonomous driving technology. Our method shows robustness in positioning. Especially, the V2X environment was constructed in RC car environment where the environment was similar to the real vehicle, and the results confirmed that the V2X environment looked for more accurate location information. There were many limitations to experiment with real vehicles in the current V2X environment such as estimating vehicle position information only by V2V system and locating vehicle by the V2I system alone. This is because autonomous driving tests using real vehicles can cause dangerous accidents if an accident occurs due to an algorithm error or sensor error during autonomous driving. So, we had difficulty experimenting with real vehicles. Since we have built the V2X environment so far, we will conduct experiments in the V2X environment in which we have built the Path Planning related research that no more can be applied to actual vehicles. So, if we continue to do research on V2X systems, we hope to have a tremendous impact on automatic driving technology in the future.

5. References

[1] C. Lienke, C. Wissing, M. Keller, T. Nattermann, and T. Bertram 2019 Predictive driving: Fusing prediction and planning for automated highway driving. *IEEE Transactions on Intelligent Vehicles, vol. 4, no. 3*, pp. 456–467

[2] P. Wang, S. Gao, L. Li, B. Sun, S. Cheng 2019 Obstacle avoidance path planning design for autonomous driving vehicles based on an improved artificial potential field algorithm. *Energies, vol. 12, no. 12*, pp. 2342

[3] Kelsey M. Judd, Michael P. Thornton, Austin A. Richards 2019 Automotive sensing: assessing the impact of fog on LWIR, MWIR, SWIR, visible, and lidar performance. *Proc. SPIE 11002, Infrared Technology and Applications XLV, 110021F*

[4] R. Changalvala, H. Malik 2019 LiDAR Data Integrity Verification for Autonomous Vehicle. *IEEE Access* pp.138018 - 138031

[5] J Wang, Y Shao, R Yu 2019 A survey of vehicle to everything(v2x) testing. *Sensors*, *19*, 334
[6] D. Elliott, W. Keen, and L. Miao 2019 Recent advances in connected and automated vehicles. *Journal of Traffic and Transportation Engineering, vol. 6, no. 2,* pp. 109-131
[7] Julio A. Sanguesa , Javier Barrachina , Manuel Fogue , Piedad Garrido , Francisco J. Martinez , Juan-Carlos Cano , Carlos T. Calafate and Pietro Manzoni 2015 Sensing Traffic Density Combi ning V2V and V2I Wireless Communications. *Sensors, 15,* pp. 31794-31810;
[8] A. Hussein, P. Marin-Plaza, F. Garcia, and J. M. Armingol 2017 Autonomous cooperative drivi ng using v2x communications in off-road environment. *Proceedings of the IEEE International Conference on Intelligent Transportation Systems (ITSC '17)*, pp. 1–6
[9] D. Jia, D. Ngoduy 2016 Platoon based cooperative driving model with consideration of realistic inter-vehicle communication. *Transportation Research Part C*, pp. 245-264
[10] V. Milans, J. Villagr, J. Godoy, J. Sim, J. Prez, E. Onieva 2012 An intelligent V2I-based traffic management system. *IEEE Trans. Intell. Transp. Syst.*, vol. 13, no. 1, pp. 49-58
[11] J. Schulz, C. Hubmann, J. Löchner, and D. Burschka 2018 Multiple model unscented kalman filtering in dynamic bayesian networks for intention estimation and trajectory prediction *21st International Conference on Intelligent Transportation Systems (ITSC),* pp. 1467–1474
[12] Q. Lim, Y. He, U. Tan 2018 Real-Time Forward Collision Warning System Using Nested Kalman Filter for Monocular Camera *IEEE International Conference on Robotics and Biomimetics (ROBIO)* pp. 868-873
[13] Q. Lim, K. Johari, U. Tan 2019 Gaussian Process Auto Regression for vehicle center coordinates Trajectory Prediction *IEEE Region 10 Conference (TENCON)* pp. 25-30
[14] Robert L.W. & Kent L.L. 1997 Modeling and Simulation of Dynamic System. *second prentice hall*

Acknowledgments
This work was supported by the Basic Science Research Program through the National Research Foundation of Korea NRF) funded by the Ministry of Education (No. 2019R1A2C108974211)

IOP Conf. Series: Materials Science and Engineering **922** (2020) 012010 doi:10.1088/1757-899X/922/1/012010

An algorithm for automatic grasping an UHF RFID passive tag

Emidio DiGiampaolo[1] and Francesco Martinelli[2]

[1]Dep. of Industrial and Information Engineering and Economics, University of L'Aquila, Italy
[2]Dep. of Civil Engineering and Computer Science, University of Rome Tor Vergata, Italy

Abstract. This paper proposes an automatic system which allows a mobile robot to reach a tagged object in an unknown environment, exploiting the Radio Frequency IDentification (RFID) technology. It is assumed that the robot has an odometry system (e.g., wheel encoders) and an RFID reader which provides it with phase-shift measurements of the RFID signal backscattered by the tagged object. Since phases are ambiguous measurements of the tag-reader distance, a Multi-Hypothesis Extended Kalman Filter is proposed to on-line extract, by fusing phase measurements with odometry readings, the range and the bearing of the unknown tag. Once the relative position of the tag is known, a control algorithm is applied to drive the robot toward the desired destination. Numerical results are reported to illustrate the approach.

1. Introduction

Localization of tagged objects is of interest in several application domains, from robotics to inventory operations in warehouses or factories. Due to the low cost and size, the absence of batteries, together with the virtual absence of maintenance, passive UHF RFID tags have attracted the research attention as a promising candidate technological solution in many contexts. Due to the low complexity of off-the-shelf hardware, only non-cooperative localization techniques based on the received signal strength indicator (RSSI) or on the phase of the backscattered signal [1] are considered. RSS-based techniques offer poor performances, since RSS suffers from noise and has low resolution and low dynamic if using off-the-shelf hardware. Phase-based techniques, instead, suffer from narrow bandwidth of RFID signal leading to ambiguities in distance evaluation. However, in dynamic environments, where either the tag or the reader is moving, it is possible to collect multiple phase measurements to overcome the lack of bandwidth [1]: this is the kind of RFID measurements considered in this paper.
In inventory contexts, it can be of interest, like in [2], to automatically map tagged objects on shelves in absolute coordinates. In these cases, the mobile unit performing the recognition should be always localized in the environment. In other applications, however, it could be of interest simply to localize a tagged object with respect to a mobile unit, in order, e.g., to reach and grasp it. This is the case, for example, of [3], which uses a Particle Filter based on an unwrapped transformation of the measured phase, to drive a human operator toward a tagged object. Unwrapping algorithms, however, may become not very effective under moderate noise situations, especially if the path of the robot is generic and quite complex, as it happens in the case of a wandering mobile unit.
The localization of a tagged object through RFID phase measurements can be faced through different approaches: fingerprinting, e.g., [4], localizes tagged objects by exploiting the similarity of phase readings with a database of measurements preliminary collected in the environment with respect to anchor tags; Synthetic Aperture Radar (SAR) methods like, e.g., [5] and [6], exploit a set of measurements from a given number of known locations to localize the unknown tag. A survey on possible localization approaches is given in [7].

In this paper, we propose a recursive, computationally efficient approach, based on a Multi-Hypothesis Extended Kalman Filter, which feeds a motion control algorithm to drive the robot to a desired objective. In order to reach a tagged object, only the relative position of the tag with respect to the reader on the mobile unit must be estimated. The relative position is captured by the tag-reader distance (range) and the bearing. The mobile unit is equipped with an odometry system (e.g., encoder on the wheels), but the robot does not have the necessity of localizing itself in an absolute frame. The key-features of the proposed strategy are, mainly, the online nature of the estimation (i.e. a range and bearing estimation is available from the first steps of the algorithm and is continuously updated as new measurements arrive, allowing the robot to immediately move toward the objective) and the fact that any robot path is suitable for the estimation (unlike other approaches, like, e.g., [2]), except possibly for straight paths, which may not allow in some situations to resolve ambiguities.

2. Notation and problem formulation

We consider a unicycle mobile robot with the pose denoted by $x_r = (x, y, \theta)$, where (x, y) is the position and θ the robot orientation. The dynamics of the robot pose can be written as:

$$\dot{x} = v_1 \cos\theta, \qquad \dot{y} = v_1 \sin\theta, \qquad \dot{\theta} = v_2 \tag{1}$$

where v_1 and v_2 are, respectively, the linear and angular velocity of the robot and can be evaluated through, e.g., the encoders mounted on the actuated wheels of the unicycle.

A RFID tag is located in an unknown position of the environment and the robot, equipped with an RFID reader and antenna, has the objective of localizing and reaching it. The unique external measurement available to the robot is the phase-shift of the RFID signal backscattered by the tag:

$$\varphi = \mathrm{mod}\left(-2K\rho + \varphi_o + n_\varphi,\ 2\pi\right) \tag{2}$$

where $K = 2\pi/\lambda$ (with λ the wavelength of the electromagnetic signal), ρ is the (unknown) tag-reader distance, ϕ_o is an unknown offset and n_ϕ is a 0-mean Gaussian noise, with standard deviation $\sigma_\phi = 10^\circ$ in our numerical investigations. The 2 multiplying K in (2) depends on the roundtrip travel of the electromagnetic wave. For simplicity, ϕ_o will be assumed 0 but the proposed algorithm can be easily adapted to handle the estimation of an unknown offset: by including ϕ_o in the state of the EKF instances, the performance of the approach is not significantly different from the case $\phi_o = 0$.

3. Solution approach

To reach the tagged object, the robot has, first, to estimate the position of the tag and, then, to apply a control strategy which drives it toward the objective. If (x_T, y_T) denotes the position of the tagged object, assumed known for a moment, we adopt the control law in [8], which assigns

$$v_1 = K_{v1}\,\rho\cos(\beta), \qquad v_2 = -K_{v2}\beta, \tag{3}$$

where K_{v1} and K_{v2} are two positive constants, while $\rho = \sqrt{(x - x_T)^2 + (y - y_T)^2}$ and $\beta = \theta - atan2(y_T - y, x_T - x)$ are, respectively, the range and the bearing of the tag with respect to the robot, (see figure 1). This control law is proved to ensure the convergence of the unicycle to the tagged object (see [8]) for any positive value of K_{v1} and K_{v2}, even if the value of these constants strongly characterizes the transient behavior of the motion.

Since the tag (and the robot) position is unknown, the range and the bearing of the tag must be estimated. The control law (3) will then be applied on the estimates $\hat{\rho}$ and $\hat{\beta}$, i.e.

$$v_1 = K_{v1}\,\hat{\rho}\cos(\hat{\beta}), \qquad v_2 = -K_{v2}\hat{\beta}. \tag{4}$$

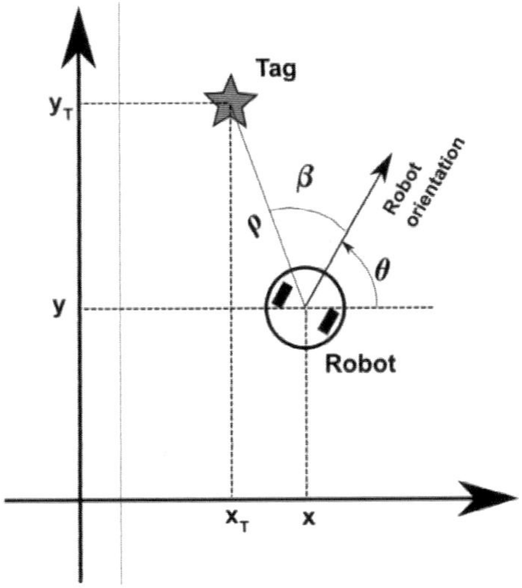

Figure 1. The robot, located in position (x, y), with orientation θ, detects the tag, located in (x_T, y_T), at a distance (range) ρ with a bearing β. The bearing is assumed positive if the tag is on the right-hand side of the robot (it is negative in the figure). Hence θ +|β| = θ−β = atan2(y_T −y, x_T −x).

The estimation of ρ and β will be realized through an Extended Kalman Filter. For this reason we have to derive the dynamics of these variables. Now, from the definition of ρ, assuming that the tag is not moving, we have $\dot{\rho} = \frac{v_1}{\rho}[(x - x_T)\cos(\theta) + (y - y_T)\sin(\theta)]$, where we have used (1). The quantity [(x_T − x)cos(θ) + (y_T − y)sin(θ)] is the scalar product of vectors [x_T −x, y_T −y] and [cos(θ), sin(θ)] and is then equal to the product of the norm of these vectors with the cosine of the angle between them, which is in fact β. But the norm of [x_T −x, y_T −y] is by definition ρ. Hence we have $\dot{\rho} = -v_1\cos(\beta)$. In a similar way, it is possible to derive the dynamics of β. Overall we have:

$$\dot{\rho} = -v_1\cos(\beta), \qquad \dot{\beta} = v_2 + \frac{v_1}{\rho}\sin(\beta). \tag{5}$$

The phase measurement (2) is related to the range ρ. If this quantity was directly measured by the robot, the system was observable (as long as the robot moves over a non straight line) and an Extended Kalman Filter, initialized with the first measurement of ρ and with an arbitrary value of β, e.g. β = 0, was able to provide a reliable estimate of ρ and β. The system remains observable even if the phase is measured. However, in view of the modulus function in (2), which represents a periodic (and hence ambiguous) measure of the range ρ, the estimation becomes more difficult and an Extended Kalman Filter is not able to solve the problem. In fact, a phase measurement gives rise to a multimodal probability distribution over the ρ values: this will make the ρ estimate of the EKF converge to the best value of the cycle where the filter has been initialized. To handle this multimodal situation, one possible solution is to use a multi-hypotheses Extended Kalman filter, that is a certain number of EKF instances, each one initialized on a different cycle corresponding to the initial phase measurement. Given the spatial periodicity of the phases (about 17cm at 867MHz, taking into account the roundtrip travel of the signal) and the usual detection range of RFID antennas in indoor environments (a few meters), 10-20 EKF instances are usually enough to cover the set of all possible cycles. The EKF instances evolve in parallel and, at each time step, they are weighed to detect which one of them is more likely to be in the correct cycle. The weight of each EKF instance is computed according to the agreement between the last, say, N_s phase measurements and the ones obtained if the tag was in the position estimated by that particular instance. The overall algorithm is as follows.

Initialization. *Let k=0 and let φ_0 be the first phase measurement available. Initialize n_M EKF instances $\ell = 1, 2, \ldots, n_M$, with $\hat{\rho}_0^\ell = -\frac{1}{2K}\varphi_0 + \ell\frac{\lambda}{2}$, $\hat{\beta}_0^\ell = 0$ and covariance matrix $P_0^\ell = \begin{bmatrix}\sigma_D^2 & 0 \\ 0 & \left(\frac{\pi}{3}\right)^2\end{bmatrix}$, where $\sigma_D = \frac{\sigma_\varphi}{2K}$ is the standard deviation of the spatial counterpart of the phase measurement noise.*

Prediction step. *The prediction step is applied independently in each EKF instance ℓ, by computing the discretized version of the dynamics (5) with the estimated quantities and using the encoder readings available, to obtain the a priori estimate* $(\hat{\rho}_{k+1}^{-,\ell}, \hat{\beta}_{k+1}^{-,\ell})$ *and its covariance matrix* $P_{k+1}^{-,\ell}$.

Correction step. *Let* $\varphi_{k+1}^{\ell} = mod(-2K\hat{\rho}_{k+1}^{-,\ell}, 2\pi)$ *be the expected phase in instance ℓ. The correction step, applied independently in each EKF instance, exploits the innovation* $\varphi_{k+1} - \varphi_{k+1}^{\ell}$.

Weighing and control step. *The EKF instances are weighed according to the agreement between the last N_s phase measurements and their expected values. Take as final estimates* $\hat{\rho}_{k+1}^{\ell^*}$ *and* $\hat{\beta}_{k+1}^{\ell^*}$ *the ones provided by the EKF instance ℓ^* with the largest weight and apply the control law (4).*

4. Numerical results

The numerical results refer to a differential drive kinematics unicycle with a distance between the actuated wheels of 26cm. The wheel displacements are measured through encoders, characterized by a 0 mean Gaussian error, with variance proportional (through a factor 10^{-4}m) to the wheel displacement. The control is applied starting from step 50. Before this step, the robot moves on a random path, generated by randomly selecting the initial position and orientation of the robot and by performing random turns when the distance to the walls of the environment becomes less than a threshold. In each step, the maximum wheel displacement is forced to be not larger than 2cm per time step, by saturating the control requests on v_1 and v_2. The control law constants are $K_{v1} = 0.1s^{-1}$ and $K_{v2} = 0.2s^{-1}$. The window to evaluate the weight of the different EKF instances comprises $N_s = 50$ steps. Figure 2 reports two trajectories generated by the proposed approach: it is possible to see, in particular on the right plot of figure 2, the part of the trajectory where the control is not active.

Figure 2. The tag, located in position (1, 0), is reached by the robot, initialized in two random positions with a random orientation, in 213 (left) and 126 (right) steps.

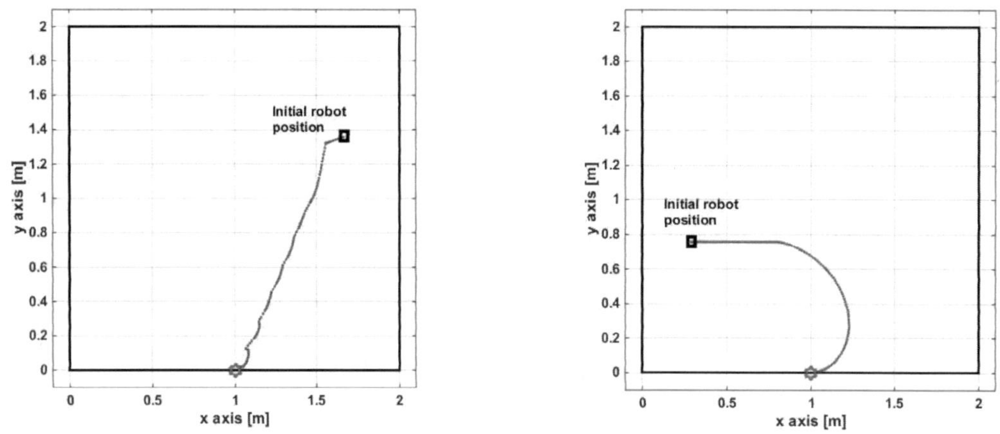

An important parameter is the frequency of measurements. Encoder readings are assumed available at all steps. The maximum displacement of the wheels in each step, as mentioned, is 2cm. Phase measurements are assumed available only every μ_s discretization steps ($\mu_s = 1$ in figure 2).
When μ_s increases, the performance deteriorates, as illustrated in figure 3, which reports, for different values of μ_s, the number of steps required to reach the tag in each execution of the algorithm over 100 different simulations (each one with a different initial robot pose and noise realization).
The performance significantly worsens starting with $\mu_s = 5$, which corresponds roughly to a sample about every 10cm. This is due to aliasing, being the spatial periodicity of the signal 17.3cm.

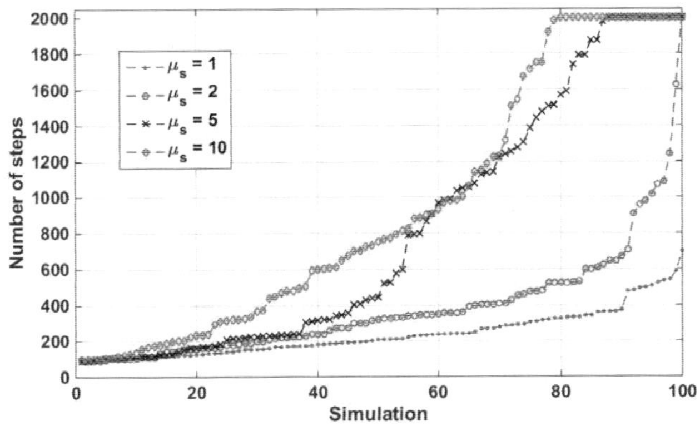

Figure 3. Number of steps necessary to reach the tag in 100 independent simulations, sorted according to the required number of steps: 2000 steps (which is the total duration of the simulations) means that the robot was not able to reach the tag in that simulation.

5. Conclusions

An automatic system to drive a mobile robot unit toward a tagged object in an unknown environment is proposed in this paper. The only external sensor used by the robot is based on the RFID technology and provides the phase-shift measurements of the RFID signal backscattered by the tagged object. The robot has also an odometry system (e.g., the wheel encoders) which allows it to know local displacements. This information is fused with phase measurements to compute an estimate of the relative position (range and bearing) of the tagged object with respect to the robot. Since the phases are ambiguous measurements of the tag-reader distance, a Multi-Hypothesis Extended Kalman Filter is necessary to solve the estimation problem. Once the relative position of the tag is known, a control motion algorithm is applied to drive the robot toward the desired destination. The proposed approach allows the robot to reach the target in an effective way, even if the interaction of the control and the estimation modules of the algorithm remains a challenging problem to be further investigated.

6. References

[1] Nikitin P V, Martinez R, Ramamurthy S, Leland H, Spiess G and Rao K V S, "Phase based spatial identification of UHF RFID tags," in *IEEE Int. Conf. RFID (2010)*, pp.102-109.

[2] DiGiampaolo E and Martinelli F, "A Robotic System for Localization of Passive UHF-RFID Tagged Objects on Shelves," *IEEE Sensors Journal*, vol. **18**, no. 20, pp. 8558-8568, Oct., 2018.

[3] Raptopoulos A, Yioultsis T and Dimitriou A G, "Particle Filter Object Tracking by a Handheld UHF RFID Reader," in *IEEE RFID-TA 2019 Symposium,* Pisa (Italy), September, 24-27, 2019.

[4] Megalou A, Tzitzis A, Siachalou S et al., "Fingerprinting Localization of RFID tags with Real-Time Performance- Assessment, using a Moving Robot," in *13th European conference on Antennas and Propagation*, Krakow, Poland, 2019.

[5] Motroni A, Nepa P, Buffi A, Tripicchio P and Unetti M, "RFID Tag Localization with UGV in Retail Applications," in *3rd International Conference on Smart and Sustainable Technologies (SpliTech)*, Split (Croatia), 26-29 June 2018.

[6] Bernardini F, Motroni A, Nepa P, Buffi A, Tripicchio P and Unetti M, "Particle Swarm Optimization in Multi-Antenna SAR-based Localization for UHF-RFID Tags, in *IEEE RFID-TA 2019 Symposium*, Pisa (Italy), September, 24-27, 2019.

[7] Li C and Mo L, "Review on UHF RFID Localization Methods," *IEEE Journal of Radio Frequency Identification*, vol. **3**(4), pp. 205-215, Dec. 2019.

[8] Canudas de Wit C, Khennouf H, Samson C and Sordalen O J, "Nonlinear control design for mobile robots," in *Recent Trends in Mobile Robots*, Y.F. Zheng, (Ed.), pp. 121-156, World Scientific Publisher, Singapore, 1993.

IOP Conf. Series: Materials Science and Engineering **922** (2020) 012011 doi:10.1088/1757-899X/922/1/012011

GPS data integration to reduce driving emissions of MHEVs in real operating conditions

Marika Fanesi and David Scaradozzi

Università Politecnica delle Marche, Italy

Abstract. Increasingly emission reduction requests and environmental pollution cut require improved optimization of current transportations. Control strategies and latest vehicles technologies developed in the automotive industry must be optimized, taking into account more information is possible from cars sensors to save energy and reduce emissions. The paper presents a possible strategy for reducing emissions with integration of GPS data in a Mild Hybrid Electric Vehicle simulator. In this study, real GPS data are evaluated to improve the system and introduce real environmental conditions in the system. The data obtained are analysed selectively in the paper. This paper is aimed to investigate the impact of real conditions on fuel consumption. Through theoretical analysis, the conclusion will be drawn that GPS technology is attractive to reduction of emissions.

1. Introduction

In recent years, the rapid automotive market development increased the number of hybrid vehicles. The hybrid vehicles, compared to other fuel vehicles, have high efficiency and excellent emissions performance[1,2]. The emissions are one of the most critical factors to take into account during the intensive study of Mild Hybrid Electric Vehicles (MHEVs) [3-6]. Last decade presented different studies of MHEVs from different points of viewsuch as modelling, systems simulation, safety, reliability, fuel consumption; all ways have been investigated to reach new and optimized solutions on environmental pollution cut. Furthermore, since the automotive field has been introduced the hybrid car, new control approaches had to be adopted [7-12]. However, the control strategies to apply to these systems are very complex in that they must consider an increasing number of variables and non-linearities. The automotive industry requests more and more real-life conditions in the simulation of the vehicle. This characteristic complicates the modelling approach and needs evidence. Nowadays, vehicles are equipped with different sensors [13-16]. Based on the data from GPS sensor and measuring the position of the car, embedded controllers could better understand the dynamical position of the car and compute the action to apply.

The precision of the new sensors distributed inside the cars has made it possible to extend the information available from the vehicle dynamics and guarantee better performance, maintaining the focus on the reducing of emissions. Also, the vehicles most considered for the production and commercialization are the MHEVs, that use two types of power sources, engine and electric motor. Previewing characteristics of the near road to face could allow deciding which type of engine claim and improve the fuel economy and emission reduction The integration of GPS sensor into the control loop could help for tracking and managing the vehicle [13], but an accurate simulator is needed to design the complete control strategy. In this paper, an MHEV is modelled and simulated. Also, respecting [3], a GPS sensor, including real data, is integrated into the system to investigate the impact of the real driving condition on fuel consumption in the simulated system. The simulator has been validated comparing behaviour on real situations versus World Harmonized Light-duty Vehicles Test Procedure cycle (WLTP) in examine the fuel consumption and emissions.

2. Proposed system design

Authors designed an improved simulator starting from the one on [3] and with the integration of the GPS information. The GPS information could be loaded from a previous survey or they could be obtained in real-time by an external serial connection. The system is designed taking into account the limitations of Government regulations and of a complete MHEV vehicle with all subsystems devoted to the motion. The overall aim of considered laws is the reduction of transport emissions and energy consumption. The benchmark levels for cars for 2025 will be 15%, and for 2030 35%. The increased focus on monitoring of 'real-drive' emission, from January 2019, justify the importance of having real data into the simulation.

2.1. Vehicle system

The system has been modelled in Matlab Simulink environment. The designed infrastructure of parallel-series MHEV includes an electrical system level with DC-DC converter and battery models, engine model with shaft and inertia and full vehicle dynamic model. The configuration of the vehicle, chosen for this simulation case, is a P2. The e-vehicle model is explained in section III of [3].

2.2. GPS technology

Recent trends in automotive industries to reach highly accurate position information and improve safety have led to introduce positioning systems. The GPS is an attractive technology to track car position and manage it. However, this technology requests a driving cycle with some features[13]:

1. Exclude short cycle;
2. Organize the direction of trajectory data;
3. Smooth the speed;
4. Divide a trip into several segments;
5. Exclude traffic influence;
6. Repeat the same route.

A GPS-recording vehicle trajectory P, with R data points, is mathematically defined in the simulator as:

$$P_R = [(p_1, t_1), (p_2, t_2), \ldots, (p_R, t_R)] \qquad (1)$$

where P represents a trajectory and p represents a data point on the trajectory with three variables: latitude, longitude and altitude; R is the total number of the data points of trajectory P. Each data point of the trajectory is recorded at timestamp t. The GPS recording interval remains the same for all trajectories of all vehicles in this study (1 second), therefor the timestamp feature is excluded from the dataset for computing reduction. A simplified data point is expressed by the following:

$$p_R = [(lat_1, long_1, alt_1), \ldots, (lat_R, long_R, alt_R)] \quad (2)$$

Supposing vehicle at the same altitude during the simulation, the Euclidian distance from p_R is calculate with the following formula where the time from one data point to the next is 1 second:

$$dist(p_t, p_{t+1}) = \sqrt{(x_{t+1} - x_t)^2 + (y_{t+1} - y_t)^2} \quad (3)$$

Where x_r is the rth latitude and y_r is the rth longitude.

2.3. Driving Cycles

Three real driving cycles, based on GPS track, was created with a real car and chosen for this study, which includes real driving data from different conditions. The three cycles has been compared with the World Harmonized Light-duty Vehicles Test Procedure cycle (WLTP), a global, harmonized standard for determining the levels of CO_2 emissions and fuel consumption of traditional and hybrid cars, used in European Country since 2015. The three real driving cycles and WLTP are summarized in Table 1. The driving cycle WLTP, Figure 1, is chosen as first to be compared with the others. The second is a Cycle in Urban Area (CUA, Figure 2), with variable speed. The third drive cycle, Cycle in

Extra-urban Area (CEA, Figure 3), is an extra-urban driving cycle that includes high-speed maintained for all life-time-cycle. The fourth, Cycle in Urban and Extra-urban Area (CUEA, Figure 4), is an urban and extra-urban driving cycle.

Table 1. Driving cycle description.

	Distance (m)	Time (s)	Average speed (km/h)	Standard Deviation
WLTP	23266	1800	46.5	36.03
CUA[a]	18820	1138	59.48	22.03
CEA[b]	47800	1470	117.06	18.69
CUEA[c]	43750	1928	81.65	25.08

[a] Cycle in Urban Area.
[b] Cycle in Extra-urban Area.
[c] Cycle in Urban and Extra-urban Area.

3. Simulation results and discussions

The GPS sensor is placed in a car with a manual shift. A car with a cruise control to maintain the same speed for a long time has been chosen. These GPS has been logged every one second. The altitude is correcting with EGM96 (Earth Gravitational Model), a geopotential model of the Earth's surface consisting of a sum of spherical harmonics up to 360 ° order and degree. The obtained GPS signal consists of a string with different information. For this study has been considered three main variables: latitude, longitude and altitude. The tests are conducted in areas with good GPS reception quality, and this has assured to provide the required signal accuracy. In this case GPS data give a good instantaneous measurement of vehicle speed. The fuel consumption and the emission in the driving cycles is shown in Table 2. The formula for the emission is described in [3]. Table 1 shows the standard deviation. The datais examined to relate the speed standard deviation of a cycle with the emission and distance. The figures 1-4 present the speed in km/h of the track and time in seconds.

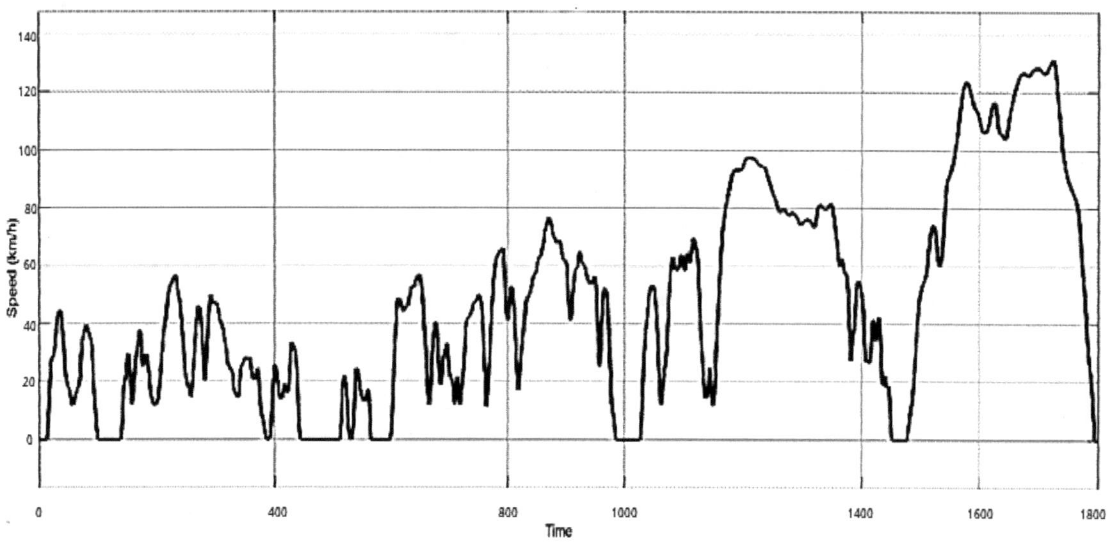

Figure 1 WLTP driving cycle

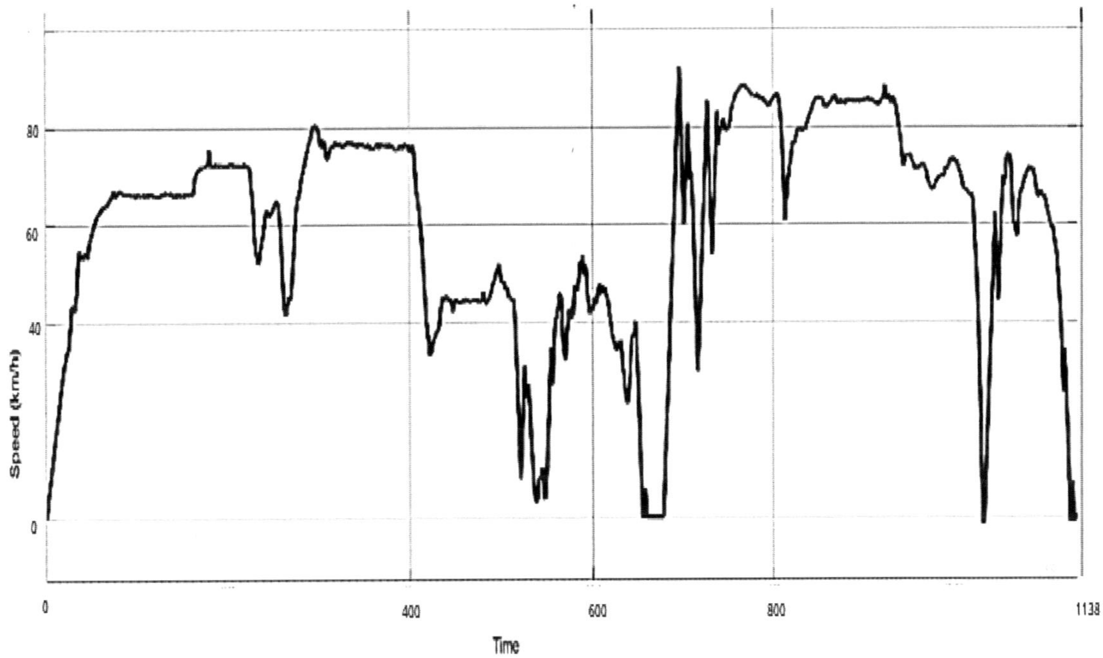

Figure 2 CUA driving cycle

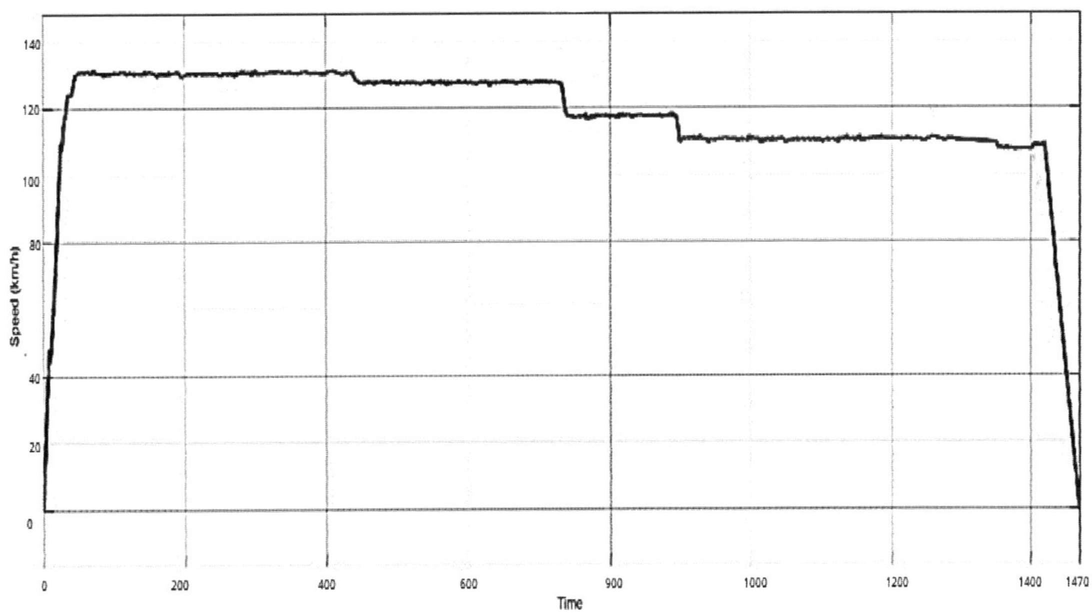

Figure 3 CEA driving cycle

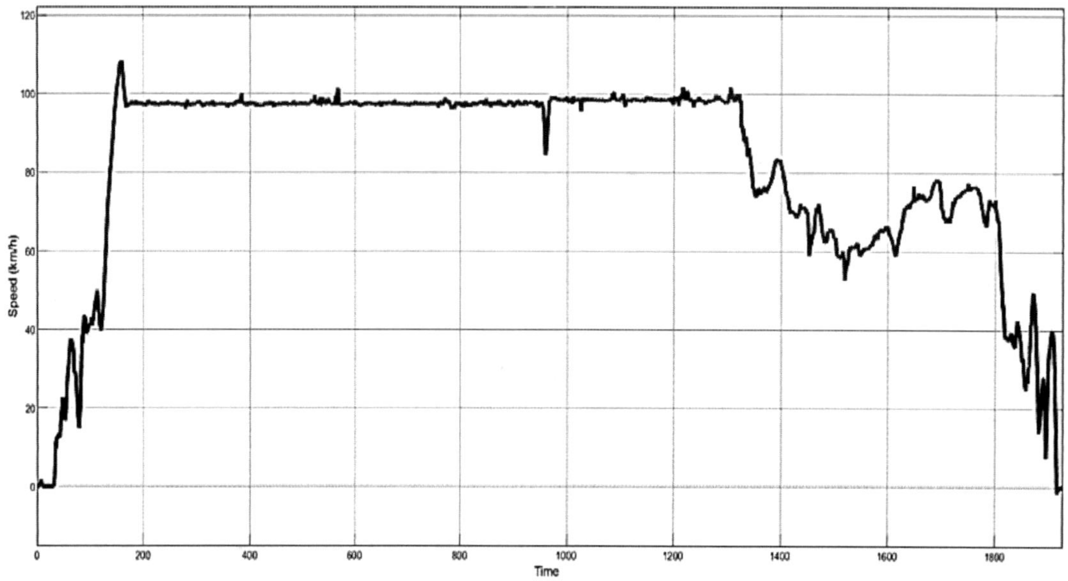

Figure 4 CUEA driving cycle

Table 2. Driving cycle emissions.

Cycle	Emission gr Co_2/km	Distance (m)	Standard Deviation
WLTP	50.95	23266	36.03
CUA[a]	62.93	18820	22.03
CEA[b]	27.62	47800	18.69
CUEA[c]	46.23	43750	25.08

4. Conclusions

This paper shows the work done integrating the GPS data into an MHEV simulator. The work presented how the GPS provides a more accurate vehicle speed profile and it allows real driving emissions analysis. The actual simulator has a PI controller inside the loop that manages the use of the electric and the traditional traction. With the improved scheme a control supervisor could be added in order to improve fuel consumption and emissions in real-time.

5. References

[1] S. M. Lukic e A. Emadi, "Effects of drivetrain hybridization on fuel economy and dynamic performance of parallel hybrid electric vehicles," IEEE Transactions on Vehicular Technology, vol. 53, n. 2, pp. 385-389, 3 2004.

[2] Q. Xiaodong, W. Qingnian e Y. YuanBin, "Power demand analysis and performance estimation for active-combination energy storage system used in hybrid electric vehicles, " IEEE Transactions on Vehicular Technology, vol. 63, n. 7, pp. 3128-3136, 1 9 2014.

[3] M. Fanesi e D. Scaradozzi, "Optimize the Mild Hybrid Electric Vehicles control system to reduce the Emission".

[4] F. Zheng, J. Li, H. J. Van Zuylen and C. Lu, "Influence of driver characteristics on emissions and fuel consumption," IET Intelligent Transport Systems, vol. 13, no. 12, pp. 1770-1779, 1 12 2019.

[5] Y. Zhang, Z. Zhang and Q. Qi, "Study on the contribution of energy saving and emission reducing of Wujiang seven cascade reservoirs' optimal regulation," in 2011 International Conference on Electronics, Communications and Control, ICECC 2011 - Proceedings, 2011.

[6] G. S. Larue, H. Malik, A. Rakotonirainy and S. Demmel, "Fuel consumption and gas emissions of an automatic transmission vehicle following simple eco-driving instructions on urban roads," IET Intelligent Transport Systems, vol. 8, no. 7, pp. 590-597, 1 11 2014.
[7] H. Yu, M. Kuang and R. McGee, "Trip-oriented energy management control strategy for plug-in hybrid electric vehicles," IEEE Transactions on Control Systems Technology, vol. 22, no. 4, pp. 1323-1336, 2014.
[8] F. U. Syed, M. L. Kuang, M. Smith, S. Okubo and H. Ying, "Fuzzy gain-scheduling proportional-integral control for improving engine power and speed behavior in a hybrid electric vehicle," IEEE Transactions on Vehicular Technology, vol. 58, no. 1, pp. 69-84, 2009.
[9] F. U. Syed, M. L. Kuang and H. Ying, "Active damping wheel-torque control system to reduce driveline oscillations in a power-split hybrid electric vehicle," IEEE Transactions on Vehicular Technology, vol. 58, no. 9, pp. 4769-4785, 11 2009.
[10] D. Kim, S. Hwang and H. Kim, "Vehicle stability enhancement of four-wheel-drive hybrid electric vehicle using rear motor control," IEEE Transactions on Vehicular Technology, vol. 57, no. 2, pp. 727-735, 3 2008.
[11] J. Chen, C. Xu, C. Wu and W. Xu, "Adaptive Fuzzy Logic Control of Fuel-Cell-Battery Hybrid Systems for Electric Vehicles," IEEE Transactions on Industrial Informatics, vol. 14, no. 1, pp. 292-300, 1 1 2018.
[12] F. Berthold, A. Ravey, B. Blunier, D. Bouquain, S. Williamson and A. Miraoui, "Design and Development of a Smart Control Strategy for Plug-In Hybrid Vehicles Including Vehicle-to-Home Functionality," IEEE Transactions on Transportation Electrification, vol. 1, no. 2, pp. 168-177, 1 8 2015.
[13] Y. Li, L. Zhao and L. R. Rilett, "Driving performances assessment based on speed variation using dedicated route truck GPS data," IEEE Access, vol. 7, 2019.
[14] Y. Wang, J. Cong, P. Wang, X. Liu and H. Tang, "A Data-Fusion Approach for Speed Estimation and Location Calibration of a Metro Train Based on Low-Cost Sensors in Smartphones," IEEE Sensors Journal, vol. 19, no. 22, pp. 10744-10752, 15 11 2019.
[15] J. Zhao, Y. Gao, Z. Yang, J. Li, Y. Feng, Z. Qin and Z. Bai, "Truck traffic speed prediction under non-recurrent congestion: Based on optimized deep learning algorithms and GPS Data," IEEE Access, vol. 7, pp. 9116-9127, 2019.
[16] X. Yun, S. Member, E. R. Bachmann, R. B. McGhee, R. H. Whalen, R. L. Roberts, R. G. Knapp, A. J. Healey and M. J. Zyda, "Testing and Evaluation of an Integrated GPS/INS System for Small AUV Navigation," 1999.

IOP Conf. Series: Materials Science and Engineering **922** (2020) 012012 doi:10.1088/1757-899X/922/1/012012

Multiphysics design and optimization of a vibration-based energy harvester from pantograph-catenary interaction

A Genovese[1], S Schiano[1], S Strano[1], M Terzo[1], M Iodice[2], M Indolfi[2] and G Coppola[2]

[1]Department of Industrial Engineering University of Naples Federico II Naples, Italy
[2]Institute of Applied Sciences and Intelligent Systems Italian National Council of Research, Naples, Italy

Abstract. In this paper, the design of a vibration-based energy harvester mounted on the railway pantograph is discussed. The system aims to feed the diagnostic system of the pantograph-catenary interaction by recovering energy from the vibrations of the pantograph head itself. The design of the system is developed through a multiphysics finite element model, parametrized by means of experimental tests. The performance of the optimized solution, in terms of generated power, highlights the affability of the proposed solution.

1. Introduction

Energy harvesting technologies have undergone an increasing interest by researchers in the last years and are nowadays used in a wide range of applications, particularly the powering of sensor nodes to monitor objects or activities.

The usage of sensor nodes become more popular also in transports with monitoring systems both for the infrastructure and the vehicles. In the railway industry, such systems are widespread because they grant a high affability and security [1],[2]. However, they are usually powered through cables, losing the advantage of mobility obtained by wireless communication, and increasing the difficulty in maintenance and accessibility. For this reason, EH appears as a valid solution for the power feeding.

The proposed work focuses on the pantograph-catenary diagnostic system for high-speed train and aims at the design of an energy harvester (EH) to feed such system, recovering energy from the vibrations of the pantograph head. Analogous systems are nowadays powered through cables that must be accurately isolated and protected due to the hostile environment. Furthermore, being the overhead contact line at high voltage, a conversion system is necessary to obtain the low voltages that drive the monitoring devices, implying significant costs and encumbrance. Subsequently, the usage of an EH would resolve these issues and, at the same time, realize a zero-cost power supply [3].

Most of the EH applications in the railway field focuses on the energy recovered from brakes [4] or on track induced vibrations [5]. Looking at train-side, some application regarding the energy harvesting from train suspensions vibration can be found [6], [7]. However, at the best of our knowledge, no applications for the pantograph-catenary system were found, leading to an innovative application in this field.

Section 2 of this paper describes the pantograph-catenary system and the studies carried out to establish the vibrations of interest. Section 3 focuses on the EH concept, while Section 4 presents the experimental tests and the modelling steps. In Section 5 the multiphysics design is carried out and the last Section summarizes the results of the study.

2. Pantograph-catenary system: analysis of the vibration source

An electric railway gathers the necessary power from an overhead wire system, called catenary, through a mechanical system mounted on the roof of the train, named pantograph.
The pantograph unfolds and extends along a vertical axis through an uplift mechanism, pushing the contact strips up against the underside of the contact wire to draw the electricity needed to run the train [8]. A good and continuous contact is then an important factor that directly affects the powertrain and is particularly difficult to guarantee at high speed; therefore, different solutions for both components have been developed [9]. Particularly, to guarantee a good contact, the contact wire must be as parallel to the ground as possible; for this reason, as schematized in Figure 1 (a), it is sustained through droppers, spaced of 6-8 m, by the messenger wire, which in turn is suspended by masts placed in intervals of around 65 m. So, the catenary stiffness is variable along the span and, due to the train travelling at a certain speed v, the pantograph head is subjected to forces and vibrations, linked to the masts and the droppers, that affect the contact quality [10].

(a)

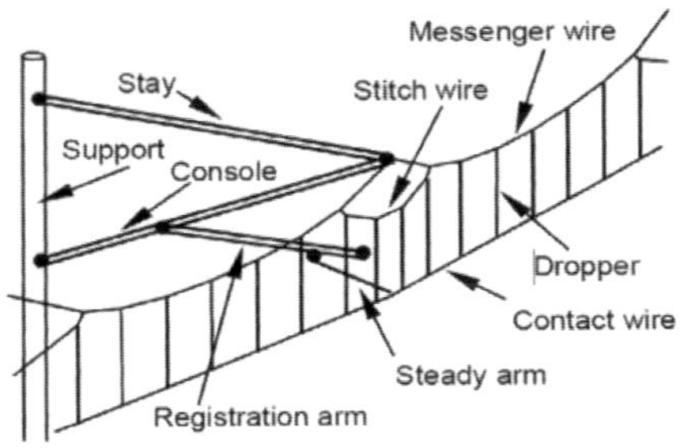

(b)

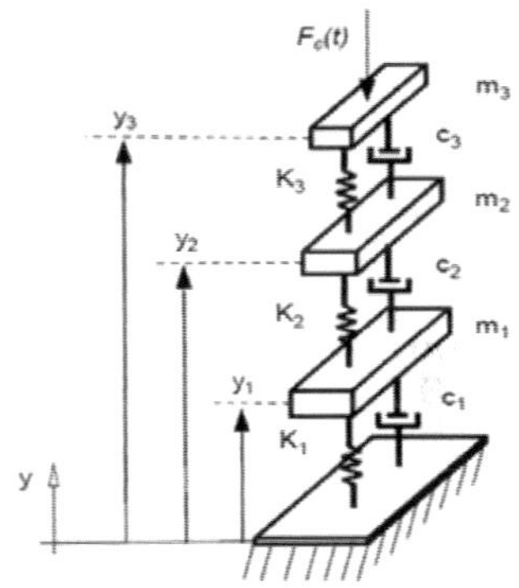

Figure 1. Representation of the overhead contact line (a) and of the 3 DOFs pantograph model (b).

The usage of active pantographs is then becoming increasingly popular; they are characterized by a diagnostic system to monitor the interaction with the overhead contact line, sending a signal to the drive system to properly modify the force exerted, guaranteeing a continuous contact.
In order to perform a quantitative analysis, the pantograph Contact ATR95, currently mounted on top of high-speed trains in Italy, was chosen for this study remarking that all in the following can be extended and scaled for any pantograph.
To establish the vibration spectrum of interest, a lumped parameter model of the current collector has been developed. Although, in the literature, models with two, three or more DOFs can be found, for high speed applications the model with three DOFs has the minimum requirements to represent the system and its dynamic response with three peaks of resonance, as found in experimental tests done to physical prototype [11]. The three DOFs model of the pantograph is represented in Figure 1 (b) and the relative motion equations are presented below:

$$\begin{cases} m_3\ddot{y}_3 + c_3(\dot{y}_3 - \dot{y}_2) + k_3(y_3 - y_2) = -m_3 g - F_c(t) \\ m_2\ddot{y}_2 + c_3(\dot{y}_2 - \dot{y}_3) + c_2(\dot{y}_2 - \dot{y}_1) + k_3(y_2 - y_3) + k_2(y_2 - y_1) = -m_2 g \\ m_1\ddot{y}_1 + c_2(\dot{y}_1 - \dot{y}_2) + c_1\dot{y}_1 + k_2(y_1 - y_2) + k_1 y_1 = -m_1 g + F_{up} \end{cases} \quad (1)$$

where y_1, y_2, and y_3 are the three DOFs. According to [11], for the pantograph under investigation the parameter set is summarized in table 1. As concerns the contact force applied on the pantograph head, F_c, which represents the interaction with the overhead contact line, it can be considered as the sum of two harmonics as in equation (2):

$$F_c(t) = F_1 \cdot \cos\left(\frac{2\pi v t}{L_s}\right) + F_2 \cdot \cos\left(\frac{2\pi v t}{L_d}\right) \quad (2)$$

where F_1 is relative to the masts and has the lower frequency and F_2 is relative to the droppers and has the higher frequency.

IOP Conf. Series: Materials Science and Engineering **922** (2020) 012012 doi:10.1088/1757-899X/922/1/012012

For the study the following operational condition has been considered: Train speed v = 300 km/h; Span between masts L_s = 60 m; distance between droppers L_d = 8 m; amplitude masts force F_1 = 150 N and amplitude droppers force F_2 =30 N. The model has been implemented in MATLAB/Simulink as a set of differential equations and the pantograph head dynamics have been evaluated. Figure 2 shows the panhead acceleration frequency response and highlights that there are two main excitations due to the interaction with the catenary, one at 1.3 Hz and one at 10.4 Hz.

Table 1. Pantograph parameters

Parameter	Description	Value	Unit
m_1, m_2, m_3	Masses	12.09, 19.41, 9.93	kg
k_1, k_2, k_3	Stiffnesses	9956.33, 19813.2, 19813.2	N/m
c_1, c_2, c_3	First DOF Damping coefficients	425.92, 42.24, 73.76	Ns/m
F_{up}	Static force exerted by the pantograph	70	N

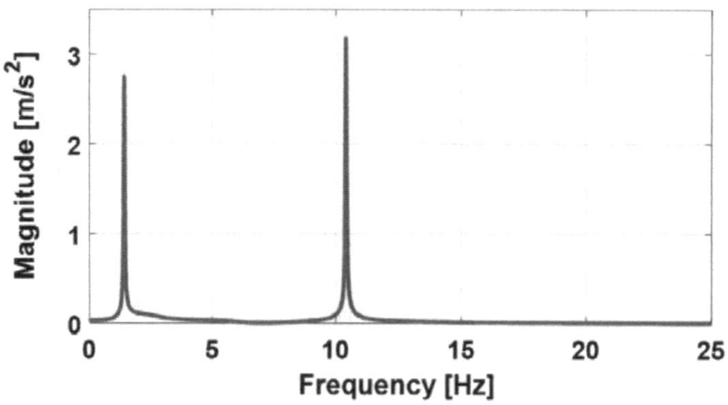

Figure 2. FFT of the pantograph head acceleration.

3. Energy harvesting system concept

Three different transducers can be used to harvest electric energy from vibrations: electromagnetic, electrostatic and piezoelectric. However, it must be considered that the railway power feeding system is characterized by high voltages both for normal and for high-speed trains (usually 3 kV/DC and 25 kV/AC respectively) [12], inducing a high electromagnetic field on the overhead contact line. Therefore, electromagnetic and electrostatic solutions could lead to severe problems. Moreover, piezoelectric transducer generally present an higher energy density [13], so it appears to be the best option.

Referring to the constitutive equations of piezoelectricity assuming a linear behaviour, normed by the IEEE Std 176 [14], the electrical energy converted by a piezoelectric system is proportional to the applied stress, so piezoelectric EH devices are designed to maximize it. For this reason, the most diffused configuration is the cantilever beam, that can be unimorph, with one piezoelectric layer, or bimorph with two piezoelectric layers. External vibrations are coupled to the cantilever through its base, causing the system to oscillate, so the piezoelectric layers undergo cycles of compression and tension generating a cyclic voltage signal across the cantilever electrodes.

A fundamental consideration in the architecture of the system is the matching between the external vibration frequency and the natural frequency of the cantilever, to maximize the deformations and thus the electrical generation. Since most vibrations can occur at low frequencies, additional proof masses are used to reduce the resonance frequency of the EH.

The concept for the implementation of the piezoelectric EH on the pantograph is presented in Figure 3, where the device is constrained between the contact strip and its housing. On this way, the excitation of the panhead is rigidly transmitted to the EH elements base.

IOP Conf. Series: Materials Science and Engineering **922** (2020) 012012 doi:10.1088/1757-899X/922/1/012012

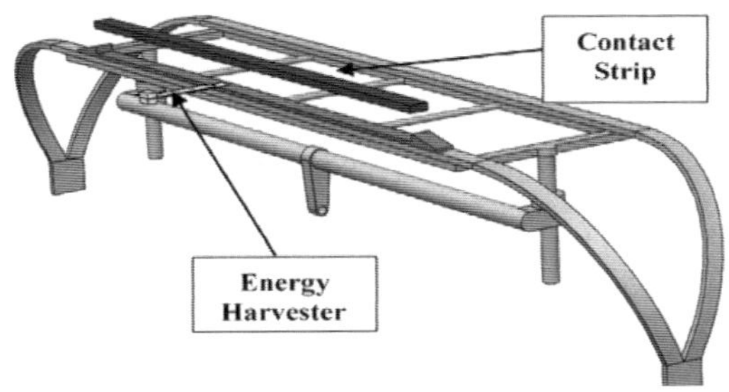

Figure 3. EH system implementation on the pantograph head

4. Multiphysics modelling

4.1. Finite element model

The piezoelectric EH behaviour has been simulated in COMSOL Multiphysics 5.4, applying the Finite Element Method. The physics that rule the phenomenon of interest are selected. Particularly, COMSOL presents a Piezoelectric Devices setup that includes the Piezoelectric Effect, Solid Mechanics and Electrostatics physics. Moreover, an external electrical circuit can be introduced and connected to the piezoelectric EH, with the aim to estimate the actual harvestable power. Of course, this power is maximal for a specific value of load impedance, which depends on the generator electromechanical characteristics and the base excitation frequency [15].

Different studies [16], [17] have demonstrated that the usage of a DC-DC converter, interleaved between a rectifier and a storage cell, solve this optimization problem: the optimal power point can be tracked by controlling the converter so that its input average current is proportional to its input voltage, with an average voltage/current ratio equal to the matching load impedance. So, a purely resistive load equal to the optimal value is connected to the PEH model, obtaining the maximum harvestable power.

4.2. Experimental model parametrization

To obtain the parameters necessary to develop a realistic finite element model, an experimental test campaign on a mock-up has been carried out. The prototype consists of eight cantilevers in unimorph configuration, composed of a steel sublayer on which is attached, with a bicomponent glue, a PZT layer. The aim is to identify the piezoelectric and the mechanical properties and parameters of the layers around the resonance to build a predictive multiphysics model.

The cantilevers are inserted between two plates and fixed on a shaker as showed in Figure 4 (a).

Figure 4 (b) depicts the experimental set-up which is composed by the EH (1); a signal generator (2) connected to an amplifier (3) that drives the shaker (4). The motions at the base of the cantilevers are measured by an accelerometer (5), while the free-end displacement of the cantilevers is measured by a laser vibrometer (6) and visualized by means an oscilloscope (8). The generated voltage is acquired by a data acquisition board connected to a voltage divider (7). A personal computer to analyse the data (9) completes the measurement chain.

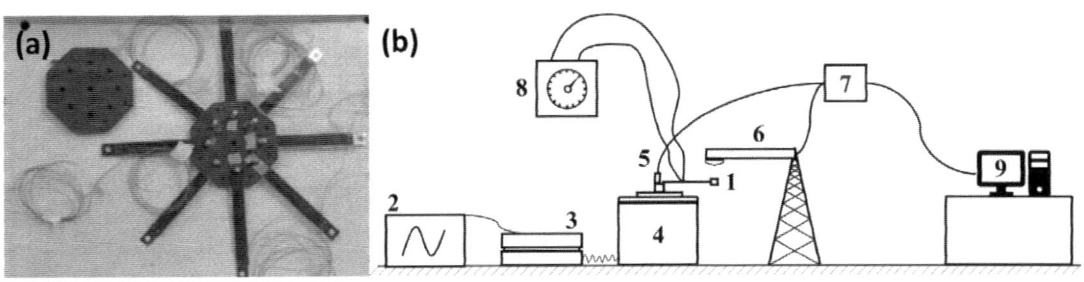

Figure 4. (a) Piezoelectric energy harvester, (b) Measurement chain scheme

IOP Conf. Series: Materials Science and Engineering **922** (2020) 012012 doi:10.1088/1757-899X/922/1/012012

The geometric and electric values of the device under test are reported in Table 2.

Table 2. Prototype's geometric and electric values

Parameter	Description	Value	Unit
l_p	Piezo length	70	mm
t_p	Piezo thickness	0.3	mm
l_i	Sublayer length	95	mm
t_i	Sublayer thickness	0.2	mm
m_p	Tip mass	34	g
w_{plate}	Energy harvester width	13	mm
R_1	Voltage divider resistance R_1	1	MΩ
R_2	Voltage divider resistance R_2	100	kΩ

The acceleration exerted through the shaker was sinusoidal with both amplitude and frequency (3-8 Hz) variable; so the ratio between the free end and fixed end displacement is considered. A finite element model is then developed with the same geometry and electric parameters of the experimental mock-up. Two studies are performed:

- *Eigenfrequency Analysis*: in order to evaluate the resonance frequencies of the piezoelectric energy harvester. This study highlights a natural frequency of the model around 5.9 Hz, i.e. in the prototype range.
- *Time Dependent*: to identify the electric power and voltage output of the harvester for a time dependent base excitation. The acquired results are compared with the experimental ones at resonance and the parameters of the numerical model were tuned until the results in terms of normalized-root-mean-square-error (NRMSE) were satisfying, both for mechanical and piezoelectrical behaviour.

Figure 5 and 6 show the comparison between experimental and simulation results at 5.9 Hz (resonance condition) for the cantilevers free-end displacements and generated voltage respectively. A very good matching between experimental and numerical results is achieved having a NRMSEs equal to 0.0061 for the displacements and 0.0098 for the generated voltage. The FEM model parameters identified at the end of the parametrization phase are summarized in table 3.

Lastly, to establish the reliability of the model, the response to varying the input acceleration frequency is evaluated and the frequency response of the generated voltage is reported in Figure 7 where the piezoelectric linear behaviour is highlighted.

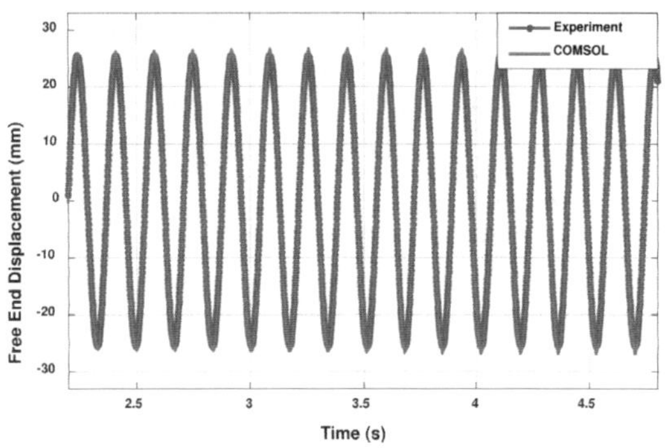

Figure 5. Comparison between the free end displacement of the EH prototype with the one obtained from COMSOL

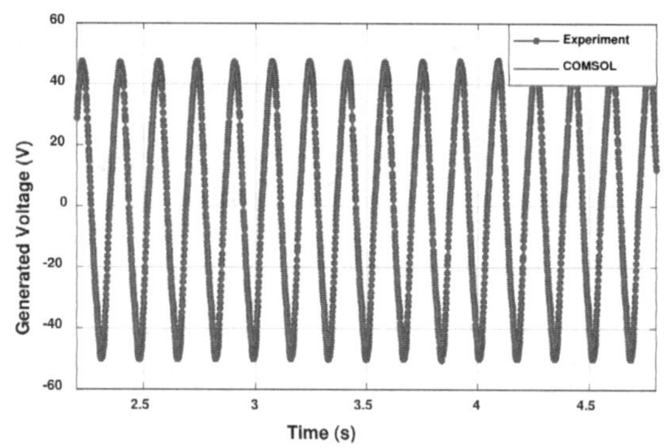

Figure 6. Comparison between the voltage generated by the EH prototype with the one obtained from COMSOL

Table 3. Results of the experimental parametrization

Parameter	Description	Value	Unit
ρ	Density	7.8E+3	kg/m^3
d_{13}	Piezoelectric Constant	-1.74E-10	m/V
ε_{11}^{T}	Dielectric Constant	1.46E-8	F/m
ε_{13}^{T}	Dielectric Constant	1.56E-8	F/m
E_{11}^{piezo}	Young's Modulus Piezo	1.590e-11	m^2/N
E_{33}^{piezo}	Young's Modulus Piezo	2.097e-11	m^2/N
$Y_{sublayer}$	Young's Modulus Steel	206	GPa
ζ_{piezo}	Damping Ratio Piezo	0.01	-
$\zeta_{sublayer}$	Damping Ratio Steel	0.04	-

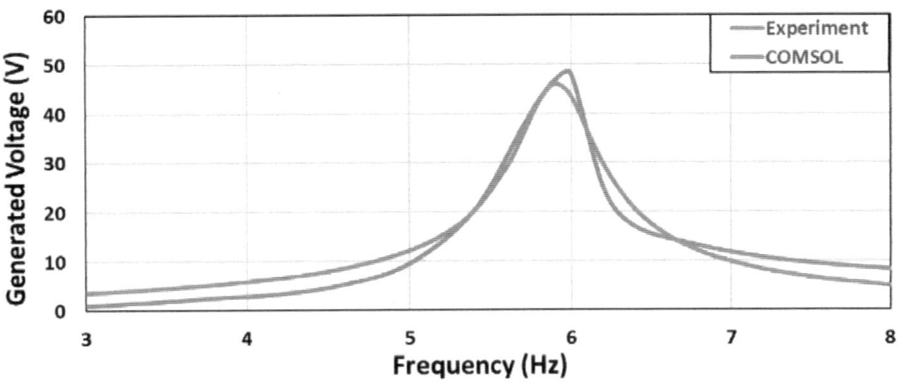

Figure 7. Comparison of the voltage generated at the vary of frequency

5. Multiphysics design

The developed FEM model has been validated so that, according to its parameters, the most suitable solution for the application of interest can be built. Considering the FFT of the pantograph head acceleration (Figure 2), the harmonic at 10 Hz is easier to match, requiring a less heavy tip mass and has a higher energy content compared to the one at 1 Hz. So, the device must have a natural frequency around 10 Hz to exploit this excitation and to work at resonance. It should be noted that, although the harvested power should be maximized, the structure can't be too heavy or bulky. Variations in the geometry and mass of the cantilever beam have then been analysed trying to satisfy design constraints and the frequency match with the excitation.

Several simulations have been performed in order find the best device layout that can guarantee the maximization of the power generation observing the design constraints, leading to a final layout consists of a bimorph structure that grants an increase in the harvested power due to the double piezoelectric layer. Adopting a geometry of 95x13x0.2 mm for the substrate and one of 70x10x0.2 mm for the piezoelectric layers, a tip mass of 16 g guarantees the match between the natural frequency of the harvester and the excitation at 10 Hz. Figure 8 represents the open-circuit voltage generated by a single piezoelectric layer, calculated as a function of the input frequency, per unitary acceleration amplitude.

Lastly, applying the pantograph head acceleration, calculated with the MATLAB model as described in Section 2, as the input excitation, the optimal load and, consequently, the maximum harvestable power has been evaluated for a parallel connection of the two piezoelectric layers (Figure 9).

About 0.3 mW can be generated from a single EH piezoelectric cantilever. Thus, considering the average length of the contact strip and the width of the proposed device, about 35 energy harvesters can be mounted for each contact strip, leading to a total power generated of 21 mW and a total added mass of approximately 1 kg.

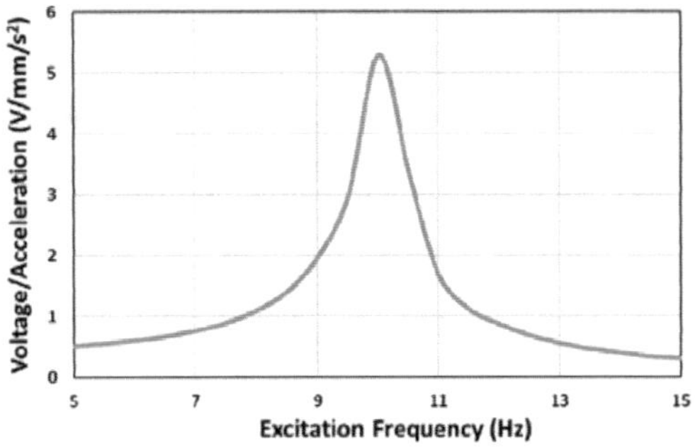

Figure 8. Open-circuit voltage generated by the bimorph PEH vs. frequency

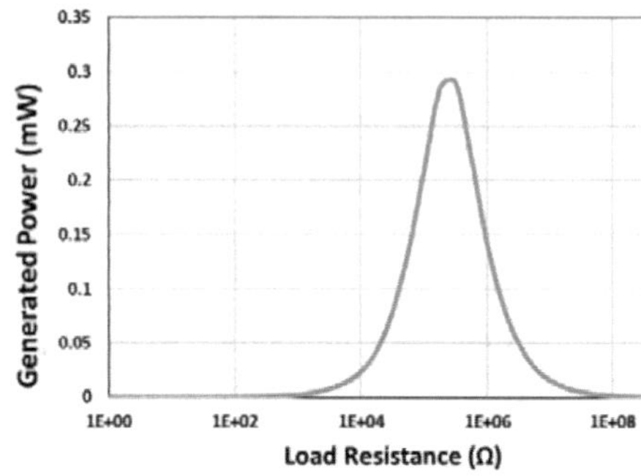

Figure 9. Power generated by the bimorph PEH vs. resistance load

6. Conclusions

The aim of this paper is the design of a piezoelectric energy harvester, mounted on the head of a railway pantograph, to recover the energy from vibrations and feed a diagnostic system, substituting the usage of electrical converters and power cables. The pantograph dynamics have been simulated through a MATLAB/Simulink model to define the forcing excitation. Then, an experimental activity has been carried out to obtain the parameters necessary to realize a consistent energy harvester finite element model in COMSOL multiphysics environment. The simulation results show an excellent match with the experimental data around the resonance. So, from these results, a modified geometry has been designed and analysed to maximize the transferable electrical power. The proposed solution appears feasible for the application of interest, yet a few considerations should be deeper examined. For example, a detailed study of the influence of the device's mass on the pantograph's dynamics should be performed to verify the mechanical interaction between the systems. Moreover, since this work was a first approach to the matter, some simplifications have been made. Particularly the interaction between the pantograph and the overhead contact line has been modelled only through the contact force, which consists of two harmonics. Yet the contact is much more complex, so it should be further investigated.

7. References

[1] C. Roberts and R. M. Goodall, "Strategies and techniques for safety and performance monitoring on railways," *IFAC Proc. Vol.*, vol. 42, no. 8, pp. 746–755, 2009, doi: 10.3182/20090630-4-es-2003.00123.

[2] "Modern techniques for condition monitoring of railway vehicle dynamics," doi: 10.1088/1742-6596/364/1/012016.

[3] A. Michael Kolbe and D.-I. Wilhelm Baldauf, "Compact contact force measurement system-online diagnosis of the overhead line system with regular trains."

[4] K. Chalkhure, S. Jijankar, A. Dhore, A. Sanganwar, V. R. Parihar, and A. Professor, "UGC Approved Journal Regenerative Braking System for Energy Harvesting from Railways and Vehicles: A Review and an Approach," *Int. J. Innov. Res. Electr. Electron. Instrum. Control Eng. ISO*, vol. 3297, pp. 2321–5526, 2007, doi: 10.17148/IJIREEICE.2017.51004.

[5] Y. Tianchen, Y. Jian, S. Ruigang, and L. Xiaowei, "Vibration energy harvesting system for railroad safety based on running vehicles," *Smart Mater. Struct.*, vol. 23, no. 12, p. 125046, Nov. 2014, doi: 10.1088/0964-1726/23/12/125046.

[6] A. Genovese, S. Strano, and M. Terzo, "Design and multi-physics optimization of an energy harvesting system integrated in a pneumatic suspension," *Mechatronics*, vol. 69, p. 102395, 2020, doi: 10.1016/j.mechatronics.2020.102395.

[7] A. Genovese, S. Strano, and M. Terzo, "Study of a Vibration-Based Piezoelectric Energy

Harvester Embedded in an Air Spring," in *5th International Forum on Research and Technologies for Society and Industry: Innovation to Shape the Future, RTSI 2019 - Proceedings*, 2019, pp. 314–319, doi: 10.1109/RTSI.2019.8895557.

[8] T. X. Wu and M. J. Brennan, "Basic analytical study of pantograph-catenary system dynamics," *Veh. Syst. Dyn.*, 1998, doi: 10.1080/00423119808969460.

[9] G. Poetsch *et al.*, "Pantograph/catenary dynamics and control," *Veh. Syst. Dyn.*, 1997, doi: 10.1080/00423119708969353.

[10] T. Wu X and M. J. Brennan, "Active vibration control of a railway pantograph," *Proc. Inst. Mech. Eng. Part F J. Rail Rapid Transit*, vol. 211, no. 2, pp. 117–130, 1997, doi: 10.1243/0954409971530969.

[11] R. M. G. A. Vieira, "High Speed Train Pantograph Models Identification Multibody Dynamics Overview," no. December, pp. 1–10, 2016.

[12] R. T. Today, "Railway Electric Power Feeding Systems," *Transport*, vol. 8, no. 10, pp. 81–83, 1998.

[13] F. Yildiz, "Potential Ambient Energy-Harvesting Sources and Techniques," *J. Technol. Stud.*, vol. 35, no. 1, Sep. 2009, doi: 10.21061/jots.v35i1.a.6.

[14] A. American and N. Standard, "IEEE Standard on Piezoelectricity: An American National Standard," *ANSI/IEEE Std 176-1987*, pp. 8–10, 1988, doi: 10.1109/IEEESTD.1988.79638.

[15] A. Koszewnik, P. Grześ, and W. Walendziuk, "Mechanical and electrical impedance matching in a piezoelectric beam for Energy Harvesting," *Eur. Phys. J. Spec. Top.*, vol. 224, no. 14–15, pp. 2719–2731, 2015, doi: 10.1140/epjst/e2015-02585-5.

[16] E. Lefeuvre, D. Audigier, C. Richard, and D. Guyomar, "Buck-boost converter for sensorless power optimization of piezoelectric energy harvester," *IEEE Trans. Power Electron.*, 2007, doi: 10.1109/TPEL.2007.904230.

[17] N. Kong, D. S. Ha, A. Erturk, and D. J. Inman, "Resistive impedance matching circuit for piezoelectric energy harvesting," in *Journal of Intelligent Material Systems and Structures*, 2010, doi: 10.1177/1045389X09357971.

AUTHOR INDEX